THE PHYSICAL WORLD OF THE GREEKS

S. SAMBURSKY

THE PHYSICAL
WORLD OF
THE GREEKS

Translated from the Hebrew by
MERTON DAGUT

Princeton University Press
Princeton, New Jersey

Published by Princeton University Press,
41 William Street, Princeton, New Jersey 08540

Copyright © 1956 by S. Sambursky

First Princeton Paperback printing, 1987

LCC

ISBN 0-691-08477-7
ISBN 0-691-02411-1 (pbk.)
Reprinted by arrangement with Routledge & Kegan Paul Ltd., Great Britain

Clothbound editions of Princeton University Press books are printed on acid-free paper, and binding materials are chosen for strength and durability. Paperbacks, while satisfactory for personal collections, are not usually suitable for library rebinding.
Printed in the United States of America by Princeton University Press, Princeton, New Jersey

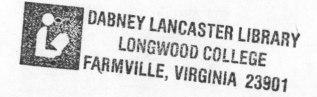

CONTENTS

v

ACKNOWLEDGEMENTS

For several of the sources quoted, available English translations have been used in this work. The most important of these have been taken from:

Ancilla to the Pre-Socratic Philosophers, translations by K. Freeman (Blackwell)

Philebus, translation by B. Jowett; *Republic*, translation by F. M. Cornford; *Epinomis*, translation by J. Harward (Oxford University Press)

The Works of Aristotle translated into English, edited by W. D. Ross (Oxford University Press)

Aristotle, On the Heavens, translation by W. K. C. Guthrie (Loeb Classical Library)

Greek Astronomy, translations by Sir Thomas L. Heath (Dent & Sons)

Lucretius, The Nature of the Universe, translation by R. E. Latham (Penguin Books)

PREFACE

THIS book, apart from some minor changes, is a translation from the Hebrew edition which was published in 1954 by the Bialik Institute, Jerusalem. It is not a history of science in Ancient Greece. As a physicist, I have been especially interested in the way the Greeks saw and interpreted the physical world around them. The texts I selected and translated in the course of my studies formed a frame into which these chapters were fitted as a kind of commentary or marginal note.

To the scientist of to-day, Greek science as revealed by these sources presents a fascinating picture: on the one hand, there is a striking similarity between its patterns of thought and ours in all that concerns scientific associations and inferences, the construction of analogies and models, and the analysis of the epistemological background. On the other hand, while we attempt to transform the world into an abstract mathematical entity which transgresses the boundaries of the inorganic universe and infiltrates into biology and the realm of man, the Greeks saw the cosmos as a living organism, as a projection of man into the distances of the outer world.

It is mainly for this reason that they were unable to visualize themselves in the Archimedean position outside the cosmos and to regard it as an object of analysis. This analysis (or dissection of nature, as Bacon called it) necessarily must proceed by "unnatural" methods such as systematic experimentation and mathematization of physical concepts. Thus the Greeks, while originating the scientific approach and thereby laying the foundation of our cosmos, were prevented, for a period of a thousand years, from making the rapid progress that came about in a few decades of the seventeenth century. From this time on, a picture of the cosmos evolved that must be set against the background of a civilization based on an interplay of science and technology, while the cosmos of the Greeks emerged from a world whose scientific curiosity remained untouched by any desire to conquer nature.

S. S.

NOTE

Numbers in brackets in the text refer to
the List of Sources quoted at the end of
the book.

INTRODUCTION TO THE SECOND EDITION

MORE than thirty years after the publication in 1956 of *The Physical World of the Greeks* this second edition appears with hardly any changes apart from the correction of a few printing errors which I could discover. I do not think that the reprint of this book as it stands needs any substantial alterations. Today however I would rewrite it with some changes in emphasis or judgment on the merits of this or that philosopher, although there are a number of problems which have not been discussed to their full extent, because the frame of this book was essentially limited to the classical and early Hellenistic period.

In the ensuing years I have come to realize that Greek science continued and developed also in late antiquity, and in retrospect I would say that this book is the first volume of a trilogy, followed by *Physics of the Stoics* (London 1959) and *The Physical World of Late Antiquity* (London 1962), both published by Routledge & Kegan Paul Ltd. The main aspects of Greek science which a third of a century ago aroused my enthusiasm for it have still retained their full validity. I shall here briefly summarize its principal points and contrast the achievements of the Greeks in this field with those of modern science.

First of all there is the marvellous gift of the Greeks for the construction of scientific theories by sheer deduction, and fruitful and pertinent speculation. The two classical examples are of course atomism, from Leucippus to Lucretius, and the Stoic continuum theory. Both are not restricted to the explanation of matter—their scope is wider, since they are expressions of opposite philosophical approaches to physical reality. While the idea of atoms whirling in the void finally led to the idea of their unexpected, causeless deviations from their courses and thus formed the physical basis of free will, the continuum theory was associated with strict determinism, with fate which in the later stages of religious development was identified with providence.

Another striking feature of Greek science is the paucity of systematic experimentation and the almost exclusive restriction to observation as a basis of its theories. If you want to explore Nature as it works you must do it with as little interference as possible, whereas the gist of an experiment is the creation of artificial or unnatural situations, conditions which are against the grain of Nature. The isolation of a phenomenon—the backbone of every modern experiment, any attempt to separate it from the complex of phenomena with which it is interwoven, was completely foreign to the spirit of the Greek scientist, as was also the repetition of a natural event through variation of the accompanying circumstances. Nature, like Man, is an indivisible unit which cannot be cut up without being distorted. Aristotle for instance who systematically developed the concept of motion in all its aspects, did not take into account the influence of friction or of the resistance of the medium on a moving body. On the contrary—these factors were for him part and parcel of the phenomenon of motion; his formula for "forced movement" led to the wrong proportionality between force and velocity.

While the correct result had to wait for Galileo, other short-comings and errors of classical Greek physics were already criticized and partly corrected in later antiquity, e.g. by John Philoponus. Some of Aristotle's classifications of movement, such as the characterization of the downward movement of a body under the influence of its weight as its striving for its "natural place" in the earth, can also be explained by the Greek conception of the unity of Man and Nature; Nature, like Man, strives for an end, and generally teleology played an important part in the explanation of many physical phenomena.

Among the most remarkable achievements of Aristotle was his thorough analysis of the fundamentals of space and time, including his clear recognition of the one-sidedness of the course of time—a serious anticipation of the concept of entropy and one of the remarkable chapters of ancient Greek thought which was sensibly elaborated by some of his Neoplatonic commentators. It is worth noting that the Greeks put the emphasis of their research on the understanding of how things work and not on the application of its results to the prediction of events. Their conception of reality was marked by a sharp dichotomy of heaven and earth. In heaven with its periodical motions everything can be predicted, whereas the sublunar region is dominated by uncertainties, the most pronounced one that of human fate.

The history of Greek astronomy from the Presocratic period to Hipparchos and Ptolemy is another impressive illustration of their capacity for constructing all-embracing theories, models which "save the phenomena" within the frame of a certain preconceived idea. Indeed, the tedious efforts, which lasted for centuries, to explain geometrically by ingenious combinations of eccentric circles and epicycles the workings of the planetary system by the geocentric hypothesis, confirmed Plato's prophetic words in the *Timeaus*, which hint at the unending struggle between the human mind ("Reason") attempting to discover order and form in physical reality, and the brute force of this reality ("Necessity"). Plato's formulation of the outcome of this struggle amounts to the statement that the final result, at its best, can merely be a partial victory of Reason; only "the greatest part of the things that become" can be explained satisfactorily, in other words—there will always remain an unresolved residue. These words of Plato were elaborated and amplified by Proclus in the fifth century. The planetary motions, he said, symbolize the essence of Nature which combines regularity and contingency, law and arbitrariness. Hegel, a great admirer of Proclus' dialectics, formulated Proclus' conclusion thus: "Nature is the unresolved contradiction".

To the unity of Man and Nature we must add the unity or rather the combination of rational and mystical thought, so perfectly exhibited by the Pythagorean School, which produced the first link in the long chain of Greek mathematics. Mathematics, the rational science *par excellence*, was originally based on the mystical, almost religious belief in the existence of a cosmic order, of the harmony of the universe which shows itself for instance in the prevalence of simple proportions in certain fundamental facts or data. Music can spiritually assist man in his awareness of this heavenly harmony, of the "music of the spheres".

The Greeks preferred pure research, knowledge for its own sake, to applied one, as I already mentioned before. Striving for knowledge is aiming for the Good as a final goal. There was most probably some connection between this attitude and the neglect of technology, as documented in their writings. Was this neglect the result of an instinctive fear of the negative consequences of technological achievements? It certainly showed the little value they attached to accomplishments of this kind, and I shall restrict myself to one conspicuous instance. Archimedes, the greatest scientist of antiquity, excelled in theoretical as well as in applied and technical achievements of supreme importance, and a large number of his treatises in mathematics, mechanics and hydrology has been preserved. However there is no doubt that several of his technical inventions, particularly his researches in the technique of warfare, were not written up and survived only in folklore or oral tradition.

In order to throw into relief the main characteristic features of Greek science it is useful to contrast it with modern science, i.e. science since the days of Francis Bacon, whose two well known slogans "science is the dissection of nature", and "knowledge is power", namely man's power to dominate nature, have become the program of scientific research from the seventeenth century on till today. The aggressive attitude of man towards nature, the disruption of the unity of Man and Nature, the cornerstone of the cosmic feeling of the Greeks, initiated an intertwining of science and technology. Science in the service of technology, and technology in the service of science led to an undreamed of rate of progress in both fields and a miraculous rise of man's standards of living and comfort. Technology went a long way in reducing illness and improving public health, but at the same time it increased the atrocities of war and developed a vicious technique of international terror. It is difficult to decide whether the overall balance of the beneficial and detrimental consequences of the merger of science and technology is positive or negative. But one thing is obvious: the development of physical science in the last forty or fifty years has vindicated some basic ideas and attitudes of Greek science. After the short interlude of positivism in the nineteenth century, which wanted to restrict science to "observables"—a trend successfully attacked by great physicists like Planck and Pauli—astrophysics and cosmology, and the physics of elementary particles on the other hand, brought about a much wider conception of reality, a discovery of a marvellous universe exceeding the range of our senses which has led us back to the primacy of fruitful speculation, the most admirable feature of ancient Greek science.

S. Sambursky
Jerusalem, September 1986

I

THE SCIENTIFIC APPROACH

"Wisdom and knowledge is granted unto thee."
2 CHRON. I.12

THE contemporary student of Ancient Greek science is forcibly struck by the historical affinity between that science and our own, no less than by the differences between them. Modern science—in particular the science of the physical world—goes back to the seventeenth century, and its origin is usually associated with the names of Galileo and Newton. Despite the many transformations to which it has been subject during these past three hundred and fifty years and to which it will in all probability be subject in the future, the character of modern science can be accurately and unambiguously defined. In method it is an interaction of induction and deduction, while in purpose it is an interplay of the comprehension and conquest of nature.

This combination is found most strikingly in physics, where systematic experimentation and mathematical formulation go hand in hand in assisting the advance of science. Just as the precise systematic scientific experiment is to-day inconceivable without theoretical mathematical calculations, so the mathematization of science increases with the growth of our experimental knowledge of nature. The instruments used have become more and more complicated; the methods of performing the experiment become more and more precise; and the mathematical description of a scientific theory and the application of mathematical principles to experimental phenomena take on ever more

1

abstract forms. In all these respects we see the constant perfecting of the reciprocal process of induction and deduction. The second characteristic synthesis of modern science finds expression in the interdependence of pure and applied science, of science for its own sake and technology. There is no branch of science or scientific theory which, starting from the purely theoretical investigation of nature, does not eventually contribute to the control of nature by the improvement of technology. Conversely, every hard-won improvement in the technical sphere infuses new vigour into pure science and enriches its theoretical bases. The time-lag between the discovery of a new principle and its practical application is growing constantly shorter, while the number of scientific problems raised by every technical invention is just as constantly increasing.

Modern science did not start from nothing; indeed, its first step was to shake off the legacy of Ancient Greece. Galileo challenged the dynamics of Aristotle, just as a hundred years before him Copernicus had constructed his heliocentric theory in defiance of the astronomy of Ptolemy. For the sake of accuracy we must note here that the revolt of the pioneers of modern science was not directed against the legacy of Greek science as such but against that petrification of its principles, and especially of Aristotle's teaching which was offered them by mediaeval scholasticism. It was a revolt against the blind acceptance of that barren bookishness which had completely divorced science from nature and the world of phenomena. It was not so much an attack on specific scientific opinions as a demand for a new scientific approach. Or perhaps we should call it a scientific revival, seeing that the original approach of Ancient Greek science to natural problems was direct and living, very different indeed from lifeless scholasticism. This is the point of contact between Greek science and the modern scientific renaissance; and it is from this standpoint that we may be called the scientific heirs of Ancient Greece.

But in what sense can there be any connection between two periods so different from each other in method and purpose? With very few exceptions, the Ancient Greeks throughout a period of eight hundred years made no attempt at systematic experimentation. This fundamental and decisive fact, reflecting as it does certain conditions and a certain mentality, will claim our attention in the last chapter. Its consequence was that induction was

2

limited to the systematic observation and collection of such experimental material as was offered by the study of natural phenomena. Such induction was naturally primitive in terms of the conception of modern science. Nor was the deduction of the Greeks any better, seeing that it lacked what Kant considered the characteristic *par excellence* of every true science: the mathematization of fundamental concepts and the deduction of facts from laws expressed in terms of mathematical formulae. Amongst the Greeks, the application of mathematics to scientific problems was confined to the description of certain phenomena, most of which were astronomical, with a few in the field of statics and optics. We find here some deductive proofs and a few calculations of cosmological data, such as the circumference of the earth or the earth's distance from other heavenly bodies. In its objective, too, ancient science is quite different from our own. It does not aim at the conquest and control of nature, but is motivated by purely intellectual curiosity. For this reason technology finds no place in it; and it suffers from the lack of that synthesis of pure knowledge and practical application which is the strength of modern science.

For all that, there is a scientific connection between the two periods. Indeed, we may go so far as to say that the roots of modern science are to be found in the Ancient World, since the basic principles of the scientific approach, which are still as valid as ever to-day, were discovered in Ancient Greece. To realize the full importance of this undeniable historical fact, we must remember what preceded the scientific revolution of the sixth century B.C. Greek science was not born in a vacuum, any more than any other science. From still earlier times and from various civilizations it inherited material some of which had been worked upon and some not: the prescientific myths and cosmogonies of Greece and the accumulated treasures of two thousand years of Babylonian and Egyptian science.

The Egyptians and Babylonians achieved their principal successes in astronomy and mathematics, just those fields where the Greeks, too, subsequently made their greatest advances. Astronomical observations in the course of two thousand years gave rise to an empirical knowledge of stellar movements with which the Egyptians and Babylonians were able to determine approximately the cycles of solar and lunar eclipses and to draw up a calendar which was subsequently adopted by the Greek astronomers, with corrections based on their own observations. No less important were the achievements in algebra and geometry. The high standard attained by Egyptian technology in metallurgy, mining and building is further evidence of a wealth of scientific knowledge. Without an extensive knowledge of mechanics and statics, and without a developed engineering technique, it would have been impossible to construct the pyramids, to transport the enormous obelisks from the quarries to the places of their erection or to carry out the actual work of erecting them. Every fresh discovery of Babylonian and Egyptian archaeology enhances our admiring respect for these scientific and technical achievements which reached their zenith hundreds of years before the birth of Greek science. But no uniform picture emerges from all these achievements, nor do the separate details coalesce to form a single body of scientific thought grounded in an all-inclusive philosophical doctrine. This had to wait for that scientific approach to the study of nature which was the creation of the Greeks in the sixth century.

This approach took the form of an attempt to rationalize phenomena and explain them within the framework of general hypotheses. The object aimed at was giving general validity to the experience obtained from regarding the world as a single orderly unit—a cosmos the laws of which can be discovered and expressed in scientific terms. It is thus no mere chance that these first steps in systematic scientific thought were taken as part of philosophy or that the natural sciences and philosophy remained wedded together throughout most of the history of Ancient Greece. One of the principal undertakings of early Greek philosophy was to give a rational interpretation to natural occurrences which had previously been explained by ancient mythologies. Greek science's achievement of independence through the struggle of logos against mythos is in many respects similar to the birth of modern science from the assault on petrified mediaeval scholasticism. With the study of nature set free from the control of mythological fancy, the way was opened for the development of science as an intellectual system. Similarly, the release of science from the domination of the philosophical dogmatism of the Middle Ages paved the way for the rise of modern science as an autonomous province of human culture.

To show that the modern and the ancient periods are alike in their scientific approach to the understanding of nature, we must have recourse to ancient philosophical and scientific literature, to a comparison of its methods with our own. In this there are two dangers. First, we must remember that a large part of Greek science has come down to us only at second hand. Most of the early period—the sixth and fifth century B.C.—is known to us mainly in the versions of Aristotle and the late commentators and compilers of the first centuries A.D., as is also the case with important scientific teachings of the post-Aristotelian period. It is probable that, in the course of such transmission, later concepts were projected into the theories of an earlier period. An even greater danger lies in the projection by the modern author of our ideas into those of Greek antiquity. Here the result might be even more distorting because of the essential difference in outlook, the reasons for which will occupy us in a later chapter. For all that, there is no good reason why this caution should lead us into the absurdity of completely foregoing any comparisons. The same nature and the same empirical data which provided the basic foundation of scientific thought in ancient times still serve us to-day, and where the subject of investigation is so clearly defined, it necessarily marks out definite patterns of understanding. The degree of success in comprehending reality may vary, but the direction taken by the investigation is the same. Hence a study of the similarities will, while showing us the limits of comparison, help us to understand the special characteristics of Ancient Greek science.

According to Aristotle, the dawn of systematic scientific reasoning occurred at the beginning of the sixth century B.C. at Miletus on the west coast of Asia Minor. Thales, and after him Anaximander and Anaximenes, were the first philosophers whose questions and answers display a truly scientific approach in our definition of the term. The three of them enquired about the physical substance underlying all phenomena, about the nature of that "primordial matter" of which all physical manifestations are made. They did not all give the same answer to this question. Thales and Anaximenes named a specific substance: the former chose "water" as the substratum, whereas the latter regarded "air" as the primordial matter. Anaximander, on the contrary, said that it was impossible to give this element a name. The question itself is important, but before we consider it let us see in what form it has come down to us in our sources.

Our authority for the doctrine of the Milesian philosophers is Aristotle, who expounds points from Thales' teaching as follows: "Of the first philosophers, then, most thought the principles which were of the nature of matter were the only principles of all things. That of which all things that are consist, the first from which they come to be, the last into which they are resolved (the substance remaining, but changing in its modifications), this, they say, is the element and this the principle of things, and therefore they think nothing is either generated or destroyed, since this sort of entity is always conserved. . . . For there must be some entity—either one or more than one—from which all other things come to be, it being conserved. Yet they do not all agree as to the number and the nature of these principles. Thales, the founder of this type of philosophy, says the principle is water" [1].

Aristotle asks why Thales chose water. The answer given by him is that the main reasons were biological. Every food contains water and every kind of seed is moist. Aristotle also considers that Thales may have been influenced by mythology, since in Greek mythology the Ocean is the father of all things. Or those who stress the physical reason may be right: water is the only substance which was known to man from the earliest times in its three different states—as solid, liquid and vapour—and which thus visibly embodied "the substance remaining, but changing in its modifications".

All these interpretations are necessarily no more than assumptions. But about one fundamental fact there can be no doubt: we have here before us an application of the scientific principle that a maximum of phenomena should be explained by a minimum of hypotheses. It may be regarded as a criterion of the simplicity of a theory if its succeeds in asserting the largest possible number of facts from the smallest possible number of assumptions. Every step in that direction can be taken as a scientific progress, and the ultimate objective, the Platonic ideal of every science would be to derive from a single root the sum total of its data. Although modern science obviously makes far more exacting demands as regards the synthesis of induction and mathematical deduction than its ancient forerunner, the two are the same in principle. For example, if physics were to succeed in deriving the forces of gravitation and electricity from a common root, we should regard this further simplification of the physical picture of the world as a great achievement. Such was the feeling of Newton's contemporaries when he demonstrated that his law of gravitation included Kepler's three laws. A similar simplification was achieved when it was proved that radio waves, light rays and X-rays are all electromagnetic radiation, differing only in wave length, and have basically the same physical qualities.

It was Thales who first conceived the principle of explaining the multiplicity of phenomena by a small number of hypotheses for all the various manifestations of matter. By virtue of his conjecture that there is a substance "from which all other things come to be, it being conserved", he became the father of all the various subsequent theories of matter, from Empedocles' theory of the four elements and the atomic theory of Leucippus, Democritus and Epicurus, through the alchemy of the Middle Ages, down to the chemistry and atomic physics of our own day. Thales and his followers went so far as to postulate a single, unchanging substance as the substratum of phenomena. The development of modern science indicates that they were right in this bold conjecture.

A second conception which the Milesian School bequeathed to future generations down to our time is to be found in its linking of the idea of primordial matter to the law of the conservation of matter. For if a single substance underlies all the changes in physical phenomena, growth and decay must be mere illusions and creation from nothing or annihilation of what exists impossible. If at a certain stage in the development of a scientific theory we have been confronted with the necessity of accepting a hypothesis involving a *creatio ex nihilo* we have enlarged the concept of substance in such a way that the laws of conservation have been restored. One classic example of this is the inclusion of heat in the concept of energy. This step was taken when it was shown that the law of the conservation of mechanical energy did not hold good for every system. Another well-known illustration is the modification of the law of conservation of matter into the law of conservation of energy, when it was proved that matter may change into radiation. The laws of conservation are now so essential a part of science that it is hardly conceivable that we should be able to do without them or that science should take any form which does not allow for the formulation of such laws. The possibility of such formulation is implied in the premiss which the Milesian School considered self-evident: that nature is capable of a rational explanation which reduces the number of variables and replaces some of them by constant quantities independent of time or of the particular form of a given process. Thus it was no mere accident that the existence of a primordial matter and its conservation were linked together in the teaching of Thales.

About Anaximander's views we are told: "Anaximander. . . said the first principle and Element is the Non-Limited. He was the first to introduce this term for the first principle. He says neither water nor anything else among the suggested Elements was first principle of anything, but there is some other non-limited substance from which all the heavens and the worlds contained in them came into being. The source from which existing things derive their existence is also that to which they return at their destruction, according to necessity; for they give justice and make reparation to one another for their injustice, according to the arrangement of Time" [5].

Anaximander did not think fit to give the primordial matter a name, since any specification would necessarily have deprived it of the essential characteristic which was the reason for its conception: namely, its complete lack of attributes. Anaximander's words show us how his opinions are to be understood. He saw that the cyclical changes of nature bring into being and destroy certain opposed qualities. As the seasons of the year change, the fundamental opposites—cold and hot, wet and dry—give place to each other. These are not abstract opposites, but are attached to certain physical states. The ascendancy of the one gives rise to "injustice", but the course of time puts this right by giving the opposite its turn to be in the ascendancy. None of the opposites can achieve absolute domination and annihilate the others. In this respect they are all limited, i.e. finite in place and time. But they are all products of the unlimited primordial matter which is an infinite reservoir of those never-ending mutations. From this observation of the cyclical changes Anaximander came to the following conclusion: "The unlimited comprises the whole cause of the generation of the world and its corruption. From it

8

the heavens were separated off and all the innumerable worlds"
[6]. This unlimited substance is not only devoid of any specific
qualities but is also the substratum of all physical phenomena and
their mutations. In the language of modern science we should
say that it is the origin of all the physical quantities, whether
mass or energy in all its forms, whether electric charge or nuclear
and gravitational forces. In the words of one of the later commentators: "He found the origin of things not in the change of
matter, but in the separation of the opposites (from the unlimited)
by means of an unending motion" [5]. These separated opposites
are the physical qualities which can be physically defined and
which, being specific, are therefore limited. The unlimited, on
the contrary, is the ultimate entity, unanalysable and indivisible:
any attempt to give it a specific name or mark of identification
transposes it into the world of specific concepts.

The third Milesian philosopher, Anaximenes, seems at first
sight rather disappointing. On closer inspection, however, we
find that he takes a most important step forward in another
aspect of the scientific approach. "Anaximenes son of Euristratus
of Miletus said, like his colleague Anaximander, that the first
matter is one and unlimited. But, unlike Anaximander, he did
not think that it is not specific, but that it is specific, namely air.
It differs in different things according to its rarity or density. In
its rare form it gives rise to fire and in its dense form it gives
rise to wind from which come clouds and water, from which
in turn comes earth, and from this stones, and from them
everything else" [12]. In other words, Anaximenes returned to
Thales' position with regard to primordial matter. He too postulates a substance with a specific quality, though unlimited in
quantity, i.e. infinite. The air of Anaximenes and of many
subsequent thinkers is undistinguishable from vapour, wind or
breath; and the transition from breath to spirit, i.e. soul, is well
known. This aspect of Anaximenes' teaching is important for the
later development of Greek physics. The Greek word "pneuma"
means breath, or air, or spirit. In the physics of the Stoics it
developed a technical meaning only remotely connected with its
original uses: by them it was primarily used to designate the
quality of the cosmos which unites all its parts into a single
organic whole through an all-pervading state of tension. This
semantic development was no doubt assisted by the dynamic

qualities of air, as they are revealed both in its elasticity and in its motion as wind. But there is also a connection with the original meaning of breath, through the analogy of the cosmos to the living and breathing creature. As long as man has breath in his nostrils, his body and all its parts form a single organic whole. Similarly the pneuma, as the breath of the cosmos, maintains the cohesion of its parts. The origin of this Stoic doctrine is to be found in the teaching of Anaximenes. There is extant a sentence which gives his own words on the subject: "As our soul, being air, holds us together, so do breath and air surround the whole universe" [13].

A further reason for identifying air with the primordial matter is that it acts as a mean between the opposites. Sometimes air takes on a cold form, sometimes a hot. At all events, it stands somewhere between fire and water. This point belongs to the other side of Anaximenes' teaching quoted above. Air (and likewise vapour which was regarded as air) appears in different forms —as hot air or cold air, as the wind which brings up the clouds, becomes dense and turns into rain. From this it is possible to generalize that matter in all its manifestations, including even all kinds of solids, is simply air in different forms. In other words, here was a way of explaining in physical terms how the various forms of matter are derived from the primordial matter. By saying "in different things it is different according to its rarity or density", Anaximenes explained the quality of the various kinds of matter by the quantity of the primordial matter: everything is air in varying degrees of density. At one end of the density scale stands fire, which is simply very rarefied air; while at the other we find all the various kinds of earth, i.e. all the solids. In the middle is water. Parallel to the density scale there is a scale of temperatures. Anaximenes was apparently also led to the identification of air with the primordial matter by the fact that air is not visible in its "normal" state, and further because its motion sometimes produces transitions from one state to another ("wind—clouds—water"). For all these reasons it occupies a middle position between hot and cold. "In form air is as follows: as long as it remains at its most average it is invisible; it becomes visible in heat and cold, in moisture, and in movement. It is in perpetual motion, since without motion it would not change" [14].

Motion was accepted by the Milesian philosophers as a basic fact which does not need explanation. In this they were followed by the later atomic school. "Everything is created by the densification and rarefication of air, but motion is from eternity" [15]. It is motion which brings about the realization of the principle that quality can be reduced to quantity. With this principle the Milesian School reached the summit of its achievement. For in this brief sentence it comprised the very essence of science from the time of Anaximenes to the present day. Throughout the ancient period this tendency was never allowed to develop fully, for the reasons already mentioned. But the mathematization of modern science has proceeded so far along this road that it has abstracted all qualities from the physical world and its phenomena and replaced them by quantity, i.e. number and measure. Although it is a far cry from the speculative teaching of Anaximenes to the extremely abstract calculations of the physicist and mathematician of to-day, yet in method both are the same. It is the method so clearly and deliberately defined by Aristotle in his *Metaphysics*: "As the mathematician investigates abstractions (for before beginning his investigation he strips off all the sensible qualities, e.g. weight and lightness, hardness and its contrary, and also heat and cold and the other sensible contrarieties, and leaves only the quantitative and continuous, sometimes in one, sometimes in two, sometimes in three dimensions, and the attributes of these *qua* quantitative and continuous, and does not consider them in any other respect, and examines the relative positions of some and the attributes of these, and the commensurabilities and incommensurabilities of others, and the ratios of others; but yet we posit one and the same science of all these things—geometry)—the same is true with regard to being" [169].

The questions which occupy the philosophers of Miletus represent the essentials of the problems which constantly recur in the physical doctrine of subsequent thinkers. Apart from speculations about the nature of matter, attention was focused on cosmology—on such questions as the structure of the cosmos, the shape of the earth, the nature of the heavenly bodies and their movements. Thales' forecast of the solar eclipse of 585 B.C. shows that he knew the empirical rules whereby the peoples of the Ancient East determined the cycles of eclipses. Herodotus tells us:

11

"It happened in the sixth year of the war [between Lydia and Media] that the battle having been joined, the day suddenly turned into night. This eclipse had been foretold to the Ionians by Thales of Miletus who had placed it within the year in which it actually occurred" [2]. Most of the conclusions drawn by Thales from astronomical observations had already been known to the Babylonians and Egyptians. His real importance lay in passing on their tradition of observation and astronomical calculation to Greece.

The spherical shape of the earth remained unknown to the Milesian School. Thales and Anaximenes assumed that the earth was flat and rested on the primordial matter. "Others say that it rests on water. This is the most ancient explanation which has come down to us, and is attributed to Thales of Miletus. It supposes that the earth is at rest because it can float like wood and similar substances, whose nature it is to rest upon water, though none of them could rest on air" [3]. This provided a "natural" explanation for earthquakes. "Thales said that the earth is carried by the water and moves like a ship. It is because of the movement of the water that the earth moves in what is called an earthquake" [4].

Anaximenes assumed that the flat earth was supported by air. Possibly he arrived at this opinion from observing that the resistance of air to falling bodies increases appreciably with their surface. In opposition to these views, Anaximander was the first to maintain that the earth is suspended in space. As regards its shape, he said: "The earth has the shape of a cylinder, the depth of which is one-third its width" [6]—that is to say, he supposed that we are on the upper face of this cylinder, while the cylinder itself rests in the centre of the cosmos. "There are some who name its 'indifference' as the cause of its remaining at rest, e.g. among the early philosophers Anaximander. These urge that that which is situated at the centre and is equably related to the extremes has no impulse to move in one direction—either upwards or downwards or sideways—rather than in another; and since it is impossible for it to accomplish movement in opposite directions at once, it necessarily remains at rest" [7].

These words are preserved for us by Aristotle, who strongly criticizes them. He compares the argument to saying a man would die of hunger and thirst if surrounded by food and drink at

an equal distance from himself. Aristotle took issue with Anaximander's theory because it conflicted with his own doctrine about
the movement of bodies to their natural place: fire, for example,
which rises upwards, does not remain at the centre, but withdraws from it. We to-day reject any *a priori* proof of a physical
doctrine which is constructed upon such considerations of
symmetry without a previous thorough analysis of the empirical
data. Nevertheless, we cannot completely dismiss the principle
underlying Anaximander's theory, as Aristotle does. In fact, this
principle has served various branches of the exact sciences, both
in ancient and in modern times. It appears in modern science as
the principle of the lack of a sufficient reason. Sometimes, when
we cannot see any adequate reason for any divergence from a
given situation, it is just this lack which seems to us sufficient
reason for the stability of that situation. In statics this principle
has served as proof of the equilibrium of a perfectly symmetrical
system. There is no reason for a lever to depart from its state of
equilibrium as long as it is symmetrical both as regards the
dimensions (length of the arms) and as regards the forces and the
direction in which they are acting. Laplace applies the same
explanation to the possibilities of a throw of a symmetrical die.
In his *Theory of Probabilities* he explicitly states that there is no
reason to assume that one case is more probable than another,
and therefore each one of the six cases must be regarded as having
the same degree of probability. Thus here too, as in the case of
the lever, we have the principle of lack of a sufficient cause
applied to a case of symmetry where Anaximander's "condition
of indifference" is operative. In principle there is no difference
between Anaximander's method and Laplace's. To-day, however,
we are more cautious in the use of this principle. We regard the
apparent absence of a sufficient cause as a sign that our knowledge
is incomplete. In the case of the dice, we presume that the centre
of gravity coincides with the geometrical centre. But Anaximander's astronomical knowledge was so inadequate that we may
say he applied a correct principle to an unsuitable object. As a
matter of fact, we employ the principle of lack of a sufficient
cause only in circumstances which make it practically the same
as its opposite, the principle of the sufficient cause. Wherever our
knowledge is imperfect, its use is extremely dubious.

In the cosmology of Anaximander use was made for the first

time of the scientific model as a means of description or as a method of explaining phenomena. This marks the beginning of the development which has culminated in the modern globe and planetarium. Anaximander was the first to "conceive the notion of drawing the inhabited earth on a tablet. After him improvements were made by Hecataeus the much travelled Milesian, until the thing became a marvel" [8]. According to another source, "he was the first to draw the circumference of land and water and also to construct a celestial sphere" [9].

Here we have the scientific model in its purely descriptive sense—the reduction in scale of cosmic dimensions to a size at which the whole and its parts can be conveniently studied. It was also Anaximander who used the mechanical model as a means of demonstrating a physical phenomenon. Even to-day, when absolute precision of terminology and mathematical abstraction rule the natural sciences, it would still seem impossible for science to dispense with the model as a means of giving concrete form to its ideas and methods. From time to time, when we want a concrete illustration of "how things work", we have to put aside mathematical abstractions and the absolute precision of the language of symbols and have recourse to a mechanical model. We form a conception of the action of elastic forces with the aid of springs, or we picture the structure of the molecule in the shape of balls joined together in a certain pattern in space. We explain propagation of sound or electromagnetic radiation by employing the model of water waves in a pool, and at present we are striving to find a suitable model for what happens in the nucleus of the atom. According to the circumstances of the case, we consider the model either as an approximation to reality, or as an exact replica of it, or simply as a makeshift which gives us only an elementary conception of the mechanism of the phenomenon. In all these cases, the mechanical model embodies a principle of scientific explanation, namely the principle of analogy. In an analogy one phenomenon is explained in terms of the functioning of another we are acquainted with or have got used to. As far as the subject or the field of investigation permits, we advance by such analogies. This general rule does not apply only to the use of concrete models: it holds good also for the transference of symbols or methods of calculation from one field to another, or for the coining by analogy of new concepts to replace the old, either through enlarging the latter or through giving them a more general application.

Anaximander's use of a mechanical model to illustrate the dimensions and movements of the heavenly bodies was an enormous advance on the allegories and mythological fancies current before his time. "Anaximander said that the sun is a circle twenty-eight times the size of the earth. It is like the wheel of a chariot with a hollow rim full of fire. At a certain point the fire shines out, through an opening like the nozzle of a pair of bellows. . . . An eclipse of the sun results from the closing of the opening through which the fire appears" [10]. This raises the question why this wheel is not visible to us and why we see only the opening in it through which the fire shines. We find the answer in another sentence from the same source: "According to Anaximander the stars are compressions of air in the form of fire-filled wheels and they throw out flames through openings in a certain place" [11]. The wheel is made of compressed air. Now, since water vapour was included in the concept of air, the air wheels are like mist or a distant cloud which cannot be distinguished from the sky. Hence, except at the opening in the envelope of compressed air, we see neither the fire in the hollow wheel nor the wheel itself, just as the sky behind mist or cloud is invisible to us. The movement of the sun across the heaven is simply the displacement of the opening due to the wheel's revolution. These two models—the revolving wheels and the fire appearing at the mouth of the forge—are perfect examples of technical analogy. They enable us to form some faint conception of the tremendous revolution in thought which took place in sixth-century Miletus.

The use of the mechanical analogy completes the picture of the scientific approach which distinguishes Ancient Greece from all that went before. Historically speaking, this is the decisive achievement beside which no great importance attaches to such details as the still very primitive theories about the relative distances of the stars. "Highest of all is the sun; next comes the moon; and below it the fixed stars and the planets" [11]—we are told in Anaximander's name. It would also seem that, in his opinion, the relation between these distances, i.e. between the radii of the wheels of the fixed stars and planets, the moon and the sun, was in the proportion of 9 : 18 : 27. These multiples of 3 were apparently taken from older cosmologies. We have already observed that the number 3 also appears in the relation of the width of the earth's cylinder to its height.

So far we have examined four noteworthy examples of the scientific approach that in the sixth century opened up a new era in the history of systematic thought: the tracing back of many phenomena to a few causes; the reduction of quality to quantity; the use of considerations of symmetry, and the employment of mechanical models. All these examples were taken from the teaching of the Milesian philosophers, which is remarkable for its rationalism.

Similar rationalistic elements are also found in two Greek philosophers of the fifth century, in Empedocles and still more pronounced in Anaxagoras. Empedocles, fragments of whose poem "On Nature" are extant, was philosopher and mystic, scientist and physician. The many-sidedness of his interests and of the intellectual expression which he gave to them makes it difficult to form a clear picture of the part played by him in the field that concerns us here. One thing, however, can be said with certainty—that to Empedocles we owe a vital theoretical addition to the foundations of science: the concept of the dependence of phenomena on universal forces at work in the cosmos. The Milesian philosophers never considered the question of a general cause; for them the transformations that occur, the constantly changing forms of the primordial matter and their physical movements were all ultimate data and attributes of the first element. Empedocles was the first to distinguish matter from force. The great originality of this lay in the distinction itself, and not in Empedocles' definition of force as a kind of material element, only of an active nature, in contradistinction to the matter activated by it. As far as matter is concerned, he explicitly propounded the existence of four elements—fire, air, water, earth. In so doing he turned his back on the monistic trend of the Milesian School and formed a new concept of matter which, with few exceptions, remained in force up to the development of modern chemistry. Bearing in mind that water and air had been postulated as elements by Thales and Anaximenes, and that fire played a similar, though not entirely equal, role in the teaching of Heracleitus, we shall understand the following words of

16

Aristotle: ". . . Empedocles says it of the four elements (adding a fourth—earth—to those which have been named); for these, he says, always remain and do not come to be, except that they come to be more or fewer, being aggregated into one and segregated out of one" [45].

These four elements underlie all the various qualities found in the sensory world. The qualities are compounded of these elements, but the elements themselves are not subject to change or disintegration. Aristotle, who thought that every genuine mixture involved a change in the quality of its components, disagreed with this view: "As for those who agree with Empedocles that the 'elements' of body are more than one, so that they are not transformed into one another—one may well wonder in what sense it is open to them to maintain that the 'elements' are comparable. Yet Empedocles says, 'For these are all . . . equal . . .'" [46]. This complete equality of the four elements was for long an accepted fact. Then the obvious effect of thermic processes, combined with early theories about the power of fire, gradually gave a position of pre-eminence to this element. Next, air was added as a second active element, on account of its elasticity and its closeness to the hot element. Hence in the post-Aristotelian period, and especially in Stoic physics, we find the elements divided into active—fire and air, and passive—water and earth. The four elements combine and separate, and it is these combinations and separations that constitute the processes of the physical world.

According to Empedocles, this creativeness is the result of two forces which he called "Love" and "Conflict" and which he poetically described as follows: "I shall tell of a double: at one time it increased so as to be a single One out of Many; at another time again it grew apart so as to be Many out of One. . . . And these never cease their continuous exchange, sometimes uniting under the influence of Love so that all become One, at other times again each moving apart through the hostile force of Hate. Thus in so far as they have the power to grow into One out of Many, and again, when the One grows apart and Many are formed in this sense they come into being and have no stable life; but in so far as they never cease their continuous exchange, in this sense they remain always unmoved as they follow the cyclic process" [41]. All the commentators are agreed that

Empedocles conceived these forces as physical quantities extended in space, but distinct from the four elements. It is thus a certain lack of precision to suppose that the four material elements with the two active forces together form a set of six elements. "He postulates four physical elements, fire, air, water and earth which are eternal and which undergo change in larger or smaller proportions through mixture or separation. There are also two ultimate principles, by which the others are moved, and these are love and conflict. For the elements are necessarily in constant motion. Sometimes they are mixed together by love, and at others they are separated by conflict. So that in his opinion there are six principles" [47].

Empedocles probably had a clear enough reason for introducing the two forces as the cause of occurrences in the cosmos: by the combination of these two opposite forces, attraction and repulsion, it is possible to describe the dynamics of these occurrences on a cosmic scale, by an integration, so to speak, over the whole cosmic field. He conceived the existence of the cosmos as a kind of dynamic equilibrium between the forces of attraction and repulsion, at the same time allowing for an alternation of world eras when one force or the other held the ascendancy. "And much the same may be said of the view that such is the ordinance of nature and that this must be regarded as a principle, as would seem to be the view of Empedocles when he says that the constitution of the world is of necessity such that Love and Strife alternately predominate and cause motion, while in the intermediate period of time there is a state of rest" [48]. The force of attraction brings about the commingling of the opposed elements, while the force of repulsion is the cause of their splitting up again into their original unmixed form. The sentence from Aristotle just quoted is certainly an idealization of the fluctuations in this dynamic condition. In fact, it is impossible to conceive of the cosmos as controlled by only one force: the forces must wax and wane alternately, as stated in another fragment of Empedocles' poem: "As they [Love and Hate] were formerly, so also will they be, and never, I think, shall infinite Time be emptied of these two" [42].

The quotation from Aristotle [48] is of special interest, because, side by side with the opposed forces, it mentions the concept of "necessity". This is one of the concepts employed by Ancient

18

Greek science to express the causal connections between phenomena. In Aristotle's opinion, which corresponds to our own
impression, Empedocles regarded Love and Conflict as the causes
of changes in phenomena, especially of the movement of matter
under the influence of attraction and repulsion. In that case,
Empedocles in his qualitative, poetic way was the first to postulate
the reality of causes in the physical world and to identify them
with forces. The modern physicist is amazed at the intuition
which led Empedocles to propound the simultaneous existence of
forces of attraction and repulsion. As far as our experience goes
to-day, the physicist has been obliged to introduce forces of these
two kinds. In opposition to the attractive force of gravitation there
are cosmic forces of repulsion at work bringing about the
expansion of the universe by the recession of the galaxies. In the
atomic field we know of the positive and negative electric charges
which repel those of the same kind and attract their opposites.
And in nuclear physics too we cannot dispense with the assumption of repulsive and attractive forces. The principle that "two
bodies cannot be in the same place" holds good in one form or
another for all circumstances and in all dimensions.

We have up to now primarily considered the basic features of
the scientific approach which took shape in the sixth and fifth
centuries B.C. and which, despite the many different forms
assumed by it, has remained essentially unchanged down to the
present day. One of these basic principles—that of seeing the
cosmos in terms of number and measure—will be more fully
discussed in the second chapter. Here we shall notice a few more
achievements which are part of the progress made in the fifth
century. These, too, are mainly the work of the two scientists
already mentioned—Empedocles and Anaxagoras.

Empedocles made two outstanding physical discoveries: first,
he recognized that air has the attributes of a body, and secondly
he taught that light propagates through space and requires time
for this propagation. The first discovery is described by him in
the course of an explanation of breathing. As proof he cites the
fact that if a vessel is immersed in water before the air has been
expelled from it, the water will not be able to enter it. The
vessel in question is the water-catcher: "As when a girl, playing
with the water-catcher of shining brass—when, having placed

the mouth of the pipe on her well-shaped hand she dips the vessel into the yielding substance of silvery water, still the volume of air pressing from inside on the many holes keeps out the water, until she uncovers the condensed stream [of air]. Then at once when the air flows out, the water flows in in an equal quantity" [43].

Empedocles' theory of light is described as follows by Aristotle: "Empedocles says that the light from the sun reaches the intervening space before it reaches the eye or the earth" [49]. A more precise description is given by a late commentator: "Empedocles said that light is a streaming substance which, emitted from the source of light, first reaches the region between earth and sky and from there comes to us. But we are not conscious of this movement, because of its speed" [50]. These two assumptions have been confirmed by modern science. The electromagnetic theory of light, and likewise the quantum theory, have shown that light is "a streaming substance". Moreover, in 1675 the Danish astronomer Roemer discovered that the velocity of light is finite and made the first approximate calculation of its value. A generation before him Galileo had already expressed the view that the velocity of light was finite but very great.

The description of light as a streaming substance emitted from radiating bodies raises a problem which was much discussed throughout the ancient period, namely the interaction of bodies and the structure of matter. The atomic school—which also goes back to the fifth century—and the Stoics from the third century on developed diametrically opposed views on this question. The former postulated the emission of particles and agglomerations of particles from the bodies which were themselves thought to be of granular structure. The latter, on the contrary, maintained an absolute continuum theory, postulating the propagation of physical processes from body to body in the form of waves. In the teaching of Empedocles and Anaxagoras we find the first tentative expression of these two views. Both of them, like all subsequent men of science apart from the atomists, denied the existence of a vacuum. Evidence of this is found in Aristotle (*De Caelo*) and in the following extant fragment of Empedocles: ". . . nor is there any part of the Whole that is empty or overfull" [44]. In this sentence we find an expression of the dilemma into which Empedocles put himself by trying to reconcile the non-existence of a vacuum with the granular structure of matter. This structure is described by him as being porous and the pores as being the recipients of the influence of other bodies: "Some philosophers think that the 'last' agent in the strictest sense enters in through certain pores, and so the patient suffers action. It is in this way, they assert, that we see and hear and exercise all our other senses. Moreover, according to them, things are seen through air and water and other transparent bodies, because such bodies possess pores, invisible indeed owing to their minuteness, but close-set and arranged in rows: and the more transparent the body, the more frequent and serial they suppose its pores to be. Such was the theory which some philosophers (including Empedocles) advanced in regard to the structure of certain bodies. They do not restrict it to the bodies which act and suffer action: but 'combination' too, they say, takes place 'only between bodies whose pores are in reciprocal symmetry'" [51]. Liquids which do not mix together, such as oil and water, lack this reciprocal symmetry. "He says that in general mixture results from the symmetry of the pores and therefore oil and water do not mix together. But the other liquids do, and he lists their specific mixtures" [52].

Philoponus, one of the late commentators, stresses that Empedocles pictured the pores as full of rarefied matter, such as air. In place of air he could have said aether, the existence of which had also been conjectured by Anaxagoras and which was conceived as the most rarified matter in existence. The aether was invented, so to speak, in order to maintain the non-existence of a vacuum. It was then elevated by Aristotle into a fifth element. In this form it passed down the generations, until its existence was questioned in the nineteenth century and finally rejected by the physicists of the twentieth century.

The second aspect of the theory of pores, which is in some respect complementary to the first, is the assumption that there are emanations from all bodies. "According to Empedocles all created things give off emanations. Not only from living creatures and from plants, from the earth and the sea do emanations flow unceasingly, but also from all the stones and from bronze and iron. Thus everything is consumed by the constant streaming away from it" [53]. It was on the basis of this theory of emanations and by the assumption of a symmetry between the emana-

21

tions of one body and the pores of another that Empedocles tried to explain magnetism. "Why does a magnet attract iron? Empedocles says that the iron is drawn to the magnet, because both give off emanations and because the size of the pores in the magnet corresponds to the emanations of the iron. . . . Thus, whenever the emanations of the iron approach the pores of the magnet and fit them in shape, the iron is drawn after the emanations and is attracted" [54].

Anaxagoras, like Empedocles, lived in the middle of the fifth century B.C. He was a native of Ionia and imbued with the same rationalistic spirit that characterized the Milesian School. In the time of Pericles he settled in Athens. The influence of his scientific theories can still be traced in late Hellenistic literature. Anaxagoras tried to prove the non-existence of a vacuum by experiment. He showed that air, which was sometimes identified with vacuum, possesses physical qualities. Besides repeating Empedocles' experiments with the water-catcher, he also (according to Aristotle) compressed air in wine skins and showed that the air offers resistance when the skins are stretched. Anaxagoras did not hold Empedocles' theory of pores, but postulated the absolute continuity of matter. His views on the structure of matter and on the intermingling of various kinds of matter were, therefore, entirely different from those of Empedocles. Anaxagoras approached the question biologically and considered the problem of metabolism. How can simple foods, such as bread and wine, be turned into flesh, sinews, bones and hair in the living body? The answer must be that they contain in themselves something of these products; and the same must be true of all the various seeds. In this way he arrived at his "seed theory", or, as it was subsequently called, the theory of "similarity of composition": "One must believe that there are many things of all sorts in all composite products, and the seeds of all things, which contain all kinds of shapes and colours and pleasant savours" [56].

In another fragment we find: "For in everything there is a portion of everything, . . . but each individual thing is and was most obviously that of which it contains the most" [57]. The later commentators further explained that Anaxagoras assumed the existence of innumerable elements, every one of which was contained as a seed in the smallest quantity of matter. That body which appears to us as "gold" contains primarily gold-grains, but it also comprises seeds from all the other elements in existence. This is the continuum solution of the problem of mixture. Later, in the teaching of the Stoics, it was combined with Empedocles' theory of the four elements and presented in a rather extravagant form.

In astronomy, the main Greek advances on Babylon and Egypt were made from the fourth century onwards. It is not worth while going into the details of Empedocles' conjectures in this field, except to note that he was aware of the centrifugal force acting when a body is swung in a circle. In order to explain why the earth stays in its place, he gave the illustration of a cup of water carried round in a vertical circle at the end of a cord. "Others agree with Empedocles that it is the excessive swiftness of the motion of the heaven as it swings around in a circle which prevents motion on the part of the earth. They compare it to the water in a cup, which in fact, when the cup is swung round in a circle, is prevented from falling by the same cause, although it often finds itself underneath the bronze and it is its nature to move downwards" [55].

Anaxagoras too made some use of centrifugal force in his cosmogony. This subject will be treated in a separate chapter, with special reference to his model of a whirl, some elements of which are also found in the cosmology of Descartes. Both thinkers assume that the aether—that extremely rarefied form of air which fills the universe—has a constant circular motion and draws the stars along with it, thus imparting a circular motion to them. Descartes, living as he did in the heliocentric period, placed the sun in the centre of this revolution, whereas for Anaxagoras the earth was the centre. "The sun, the moon and all the stars are flaming stones which are carried round by the revolution of the aether" [62]. In this way—by means of this simple mechanical model and the postulation of an invisible medium occupying the space between the centre and the bodies and involving them in its circular motion—the revolutions of the heavens were explained by a single mechanism. The great importance of Anaxagoras in the annals of astronomy lies in this mechanical outlook which is the logical continuation of the rational approach laid down by Anaximander. Anaxagoras' astronomical hypotheses are throughout dominated by a "terrestrial" approach which makes no distinction between phenomena "there" in the sky and those "here" on the earth, and gives a purely physical evaluation of astronomical data and their possible causes. The heavenly bodies are nothing more than flaming stones; "the sun is larger than the Peloponnesus" [62]—what uninhibited freedom of thought is revealed by this comparison of the mightiest of the celestial bodies, apotheosized by the deep-rooted irrational beliefs of mythology, with a geographical object, a part of the inhabited earth! To-day such rationalization is a commonplace. But the breach made by Anaxagoras in the fortress of mythology at one of its strongest points was an event that had few parallels in the long history of the ancient world.

Just how revolutionary his achievement was can be seen from his views on the origin of the large meteor which fell at Aegospotamoi. "It is generally believed that a huge stone fell from the sky on Aegospotamoi. The inhabitants of the Chersonese point it out to this day and pray to it. Anaxagoras is said to have held that this stone came from one of the heavenly bodies on which there was a landslide or earthquake, with the result that this stone was broken off and fell upon us. For no star remains in the place where it was created. The bodies of these heavy stones shine because of the resistance of the revolving aether and are forced to follow the whirl and pressure of the circular motion" [63].

Most of the Greek astronomers accepted Anaxagoras' theory that the moon has no light of its own, but that "it is the sun that endows the moon with its brillance" [64]. An exception was Poseidonius the Stoic (second century B.C.) who reverted to the former view that the moon emits light of its own. In later writings, there are likewise references to Anaxagoras' doctrine that "the moon has inhabited parts as well as mountains and ravines" [65[, and in Plutarch's "On the Face of the Moon" this point is discussed in detail.

While it is true that, generally speaking, Anaxagoras' astronomical ideas were primitive and not original, this in no way detracts from his importance as a pioneer of the concept of the unity of celestial and terrestrial phenomena. Through it he fell foul of the authorities, was tried on a charge of atheism, sentenced to imprisonment and only set free on the personal intervention of

Pericles. He thereupon left Athens and returned to his native land. In the history of science his place is also assured as the first thinker to stress the dependence of our physical experience upon our senses. "Through the weakness of the sense-perceptions, we cannot judge truth" [60]. He proved this statement by a simple experiment in which he showed that we cannot distinguish the slight change that occurs in a colour when another colour is added to it drop by drop. Together with his recognition of the limitations of sensory perception, he enunciated the important statement that sensation is evidence for a physical reality that cannot be comprehended by our senses. "Visible existences are a sight of the unseen" [61]. This realization became one of the basic principles of the Greek atomists. It has been turned from hypothesis into fact by modern atomic theory.

II

NATURE AND NUMBER

"And I lifted up my eyes, and behold a man with a measuring
line in his hand."   ZECH. 2.1

———

IN the first chapter we discussed the foundations of science
which were laid in the sixth and fifth centuries B.C.; we gave
examples of the first beginnings of the scientific approach as they
are revealed to us in the Milesian School's doctrines of matter and
force and in the teachings of Empedocles and Anaxagoras. Here
was the origin of that train of thought which led to the atomic
school, to alchemy and chemistry, and to our contemporary theory
of matter. But this was not the only scientific chapter to be
opened in that period. There was another of even greater importance in its scope and repercussions: the first attempt to see the
cosmos and what happens in it in terms of number and measure.
The pioneering work of Pythagoras and his school in this field
was continued by Plato and the mathematicians of the Hellenistic
era, and finally given a new significance by Galileo, Kepler and
mathematical physics from Newton to our own day.

At this point it would be as well to reiterate the importance of
the rich scientific legacy inherited by Greece from Egypt and
Babylon, especially in mathematics. Pythagoras, who spent many
years in these countries, was undoubtedly very well acquainted
with the great discoveries made by the Babylonians in arithmetic
about one thousand five hundred years before his time, and had
mastered Egyptian geometry which went back much further
still. At the same time, however, if Pythagoras and his disciples

had done no more than to add their own discoveries to this store, their work, though sure of an honoured place in the history of mathematics, would not have had the great scientific significance which is attributed to it.

Pythagoras was born in Samos, but in the second half of the sixth century he emigrated to southern Italy, where he gathered round him a group of disciples who lived a communal life devoted to the mysteries of philosophy and mathematics. More than a hundred years after his death some of his pupils began to disregard their master's command of secrecy. Thus the original oral teaching was gradually supplemented by written works, some fragments of which have survived. From these and from the commentaries of Aristotle and the later philosophers, especially the Neo-Platonists, it is possible to get to know the main outline of Pythagorean doctrine. Pythagoras was mainly concerned with the properties of integral numbers. He discovered several theorems which we should classify as belonging to arithmetic or to the elementary theory of numbers. His method was to arrange numerical sets in geometrical form: every unit was represented by a pebble which was put in the place of a physical or geometrical point. In this way it is possible to arrange the integers symmetrically in a series of rows of points one under the other. The result is a triangle with One at its apex and under it the number two in the form of two points, followed by a row of three points and so on. Add the rows together and we get the series of "triangular numbers", 1, 3, 6, 10, 15, etc. From the arrangement of the numbers in rows it can be clearly seen that every

Fig. 1. Triangular diagrams.

triangular number is equal to the sum of all the integers from one down to the serial position of the triangular number in question. For example: six, the third triangular number, is the sum of one, two and three; ten, the fourth in the series, is the sum of the integers from one to four, and so on. When we remember that the Greeks used to designate numbers by the letters of the alphabet—a system which completely conceals the laws of sequence in a series of numbers—the superiority of Pythagoras' method is at once evident. The simple fundamental fact that odd integers alternate with even is obvious at a glance from the above triangle of points, where odd and even rows appear one after the other. This method of description also showed Pythagoras' pupils that numbers and their sequences inhere in physical bodies, such as those pebbles which are the elements of more complex bodies. Unity, the single pebble, is also a body in itself; it can be extended into lines, which can be made into geometrical figures in a plane which again may bound bodies in space. From the single dimension of the unit we proceed to the higher dimensions of the line, then to the plane figure, such as a triangle, square, etc., and finally to the pyramid, cube and other bodies. Since zero was unknown to the Greeks, unity performed two functions: it was both one-dimensional unit of construction and non-dimensional point of contact between two sections. Now, the perfect bodies are simply an idealization of the physical patterns in the cosmos. Hence it may be said that number underlies all physical objects and is the beginning of everything.

This is the origin of Pythagoras' philosophy of numbers which his esoteric and mystic disposition turned into a creed. Though disapproving of this Pythagorean religion and its influence on his master Plato, Aristotle tried to describe its main tenets dispassionately: "Evidently, then, these thinkers also consider that number is the principle both as matter for things and as forming both their modifications and their permanent states, and hold that the elements of number are the even and the odd, and that of these the latter is limited, and the former unlimited; and that the One proceeds from both of these (for it is both even and odd), and number from the One; and that the whole heaven, as has been said, is numbers" [22]. We see that the principle of number lays bare a far deeper layer of reality than the principle of primordial matter could reveal. For whatever this matter may be, it always takes the form of number, in a certain geometrical or arithmetical combination which lends itself to numerical or mathematical formulation. Every substance that can be perceived by our senses appears in a certain numerical arrangement; and every physical entity, be it an actual substance or the state of a physical system or a change occurring in it, can always be expressed in terms of number or by a mathematical formula. This Pythagorean conception is the starting-point from which our modern mathematization of physical phenomena developed. In Kant's words, the province of science is whatever can be expressed in mathematical terms.

To understand the significance of even and odd numbers in Pythagoras' teaching we must refer back to that arrangement of physical points in rows which gave rise to the concept of number. Every unit is represented by a particle separated from the others by a certain interval. It is this isolation which gives rise to number, whether to the unit or to the number composed of several units. In this sense the Pythagoreans postulated the existence of empty space, not however as an absolute concept, like Leucippus and Democritus, but, following Empedocles, in a relative sense. [23]. This void may be air or aether or any "rare" medium, provided that it isolates number, or the body representing it, as a separate unit. In this conception of number as bound up in discrete particles is to be found the notion of the atomic theory according to which the cosmos is a collection of "full" physical units separated from each other by a void. For Pythagoras number was the expression of the two basic phenomena of the physical world—extension and shape. Matter which fills space by its extension is the "unlimited" and is represented by the even number, since the most obvious characteristic of even numbers is that they can be divided into equal parts. When we bisect one of the rows of Pythagoras' triangle of points we do not touch any limit, if the number of points in the row is even: the dividing line passes through the space between two points. But if the number of points is odd, the line merges with one of the points; this point then stands for the limit which produces a certain shape by turning an unlimited line into a section of definite length. Thus, in higher dimensions, the line becomes the limit of the surface, and the surface the limit of the body. Since the even and odd numbers alternate with each other to produce the sum total of the integers, number became for Pythagoras the source of both the unlimited and the limited, of the extending substance and of the shape that limits its extension. In the words of the Pythagorean philosopher Philolaus: "Nature in the universe was fitted together from the Non-Limited and Limiting, both the universe as a whole and everything in it" [16].

The fundamental quality of numbers, by virtue of which they combine the opposites of substance and form, is harmony. "This is how it is with Nature and Harmony: the Being of things is eternal, and Nature itself requires divine and not human intelligence; moreover, it would be impossible for any existing thing to be even recognized by us if there did not exist the basic Being of the things from which the universe was composed, both the Limiting and the Non-Limited. But since these Elements exist as unlike and unrelated, it would clearly be impossible for a universe to be created with them unless a harmony was added, in which way this (harmony) did come into being. Now the things which were like and related needed no harmony; but the things which were unlike and unrelated and unequally arranged are necessarily fastened together by such a harmony, through which they are destined to endure in the universe" [17]. This fragment, like most of the extant fragments of Philolaus, is filled with a religious faith in the regulating power of number and harmony within the cosmos. It would not be out of place to quote a few more of these fragments, since the spirit with which they are infused had such a profound influence on Plato:

"Actually, everything that can be known has a Number; for it is impossible to grasp anything with the mind or to recognize it without this" [18].

"The nature of Number and Harmony admits of no Falsehood; for this is unrelated to them. Falsehood and Envy belong to the nature of the Non-Limited and the Unintelligent and the Irrational. Falsehood can in no way breathe on Number; for Falsehood is inimical and hostile to its nature, whereas Truth is related to and in close natural union with the race of Number. For the nature of Number is the cause of recognition, able to give guidance and teaching to every man in what is puzzling and unknown. For none of existing things would be clear to anyone, either in themselves or in their relationship to one another, unless there existed Number and its essence. But in fact Number, fitting all things in to the soul through sense-perception, makes them recognizable and comparable with one another" [19].

This emphasis on the interconnection between number and the sensory world, as first found in the doctrine of Pythagoras, could have been the beginning of a mathematical explanation of the physical world on the lines of modern theory. But Plato distorted this development by abandoning the belief that the cosmic harmony is revealed through contact with the sensory world. The Platonic theory of "ideas" gave rise to a belief that the cosmos could be comprehended by pure mathematics and that this was only obscured by contact with matter and empirical phenomena. But, before we come to this particular question, it would be as well to survey the effect of the Pythagorean doctrine upon the character of the Greek cosmos. We shall not touch on all the mystic offshoots of the religion of numbers which, with their manifold ramifications from the Neo-Platonists onwards and throughout the Middle Ages, finally became part of Christian and Jewish mysticism. Nor shall we consider the full scope of the great impulse given by this doctrine to Greek mathematics, which in the three hundred and fifty years after Pythagoras laid the foundations of arithmetic and explored all the aspects of geometry. It is true, no doubt, that both numerical mysticism and mathematics played a part in moulding the Greek conception of the cosmos. But here we are only concerned with the direct influence exercised by the Pythagorean School upon the rational approach to the comprehension of the cosmos.

In this connection three things deserve special mention: the concept of geometrical perfection which is part of the concept of harmony; the importance of mathematical proportion; and the discovery of irrational numbers. The five perfect bodies, or Platonic bodies as they are also called, are discussed in detail by Euclid in the 13th section of his *Elements*. According to one of the later commentators of Euclid, three of these bodies were discovered by the Pythagoreans and the other two by Theaetetus, a contemporary of Plato. Four of the five were given a cosmic significance by each being associated with one of the four elements: the cube bounded by four squares was associated with earth; the tetrahedron bounded by four equilateral triangles with fire; the octahedron (8 triangles) with air; and the icosahedron (20 triangles) with water.

A blend of these Pythagorean ideas and Democritus' atomism became the basis of Plato's geometrical theory of matter as he describes it in his *Timaeus*. Here we may confine ourselves to the essentials of his doctrine. Plato tries very hard to uphold Pythagoras' teaching about the numerical harmony which imposes form on matter. Every one of the four physical elements has its own numerical mould. As matter extends in space, it follows that this mould must be linked to a spatial geometrical figure. Mathematics has conveniently provided those who sought for such figures with the perfect bodies which were soon found to be only five in number. Three of them are bounded by triangles, one by squares and the fifth (the dodecahedron) by perfect pentagons. The decisive consideration in the choice of the first four bodies was the need to find a common element in all of them. This need arose from the fact that the elements through mixture become the source of all material phenomena.

Here recourse is once more had to the generative principle of Pythagoras whereby a figure of a certain dimension is made up of elements of the preceding dimension. A line, for example, is composed of a row of points; in the same way, the faces of the perfect body and their spatial position were made to delimit the object in space. The simplest solution was to find a "common denominator" for the bodies bounded by triangles and for the cube. For if we halve a square we also get a triangle which, though not equilateral, is at least isosceles, and which makes up with its right angle for what it lacks in the perfection of its sides. The perfect triangle can likewise be divided by its perpendicular into right-angled triangles, though in a somewhat lower degree of perfection—instead of two equal sides we find here one of the sides adjacent to the right angle in the ratio of one to two to the hypotenuse. By combining together Pythagorean and Democritean ideas Plato introduced these two right-angled triangles as "atomic" elements into his theory of matter. From the combinations of these plane atoms and their spatial arrangements are created the four elements which can mix and pass into each other. True, the pentagons which bound the dodecahedron cannot be broken down into right-angled triangles. But even here a way out was found after the introduction of the aether into the cosmic picture as the finest and purest of all substances whose domain was the upper levels of the cosmos in the regions of the stars. Thus the fifth perfect body was associated with the fifth element of Aristotelian terminology.

There is a further significance to this matter of the five perfect bodies. Their study led the pupils of Pythagoras to recognize the cosmic importance of the sphere. They were the first to teach that the earth, and likewise the whole universe, is round: "It is said that he [Pythagoras] was the first to call the heavens cosmos and the earth a sphere" [24]. No doubt this discovery was in part due to the sight of the earth's round shadow passing across the moon during an eclipse. But probably no less decisive were "theoretical" considerations, especially with regard to the sphericity of the whole cosmos which contains all the physical elements, just as a sphere contains all the perfect bodies: "The bodies of the Sphere are five: the Fire in the Sphere, and the Water and Earth, and Air, and, fifth, the vehicle of the Sphere" [20]. The last words of this sentence from Philolaus appear to contradict the foregoing interpretation. But, besides a possible corruption of the text, we must bear in mind the important position that the aether had come to occupy in the course of time. In this passage the concept of aether has virtually merged into that of the sphere of the fixed stars. At all events, this was the view of Theon of Smyrna (second century A.D.), to whom we owe the above words of Philolaus. The discovery of the spherical shape of the earth was the main contribution of the Pythagorean School to astronomy. But their other cosmological conjectures also had a considerable influence upon the history of astronomy, and we shall have more to say about them.

Pythagoras found the harmony of numbers most strikingly displayed in the mathematical ratios between various numbers, taking the form either of proportions or of geometrical relations. The proportions will be discussed later in connection with musical harmony. The most famous of the geometrical theorems, which have also an algebraic significance, is that still known by the name of Pythagoras: in a right-angled triangle, the sum of the squares of the two sides is equal to the square of the hypotenuse. The Egyptians had already discovered empirically that three sections of 3, 4 and 5 units respectively in length form a right-angled triangle. But it was left to the Pythagorean School—perhaps to Pythagoras himself—to find the general proof and to discover other "Pythagorean" numbers which conform to the theorem, e.g. 5, 12, 13 or 8, 15, 17, etc. This discovery that sets of integers conform to a geometrical law together with the Pythagorean custom of representing integers by rows of pebbles, gave rise to notions which were quickly proved false. They were based on a kind of naïve theory of geometric "atomism", according to which the ratios between geometric quantities could be expressed by ratios of integers, as it was assumed that they could both be reduced to the common denominator of the atomic points contained in them. However, the disciples of Pythagoras themselves found out that elements of even the simplest and most perfect geometrical figures bear ratios to each other that cannot be expressed in these rational terms. This discovery that there are numbers "devoid of logos", irrational numbers, shook the foundations of the Pythagorean belief in the essential harmony inherent in the physical world, and for a long time it was kept secret. Later on, however, it served as a stimulus for a more profound understanding of the world of numbers and of the continuum of geometrical points corresponding to it. The existence of an irrational number was first shown by the demonstration that the lengths of the side of a square and its diagonal cannot be expressed as the ratio of two integers. From the application of Pythagoras' theorem to a right-angled isosceles triangle it follows that there are two squares whose ratio equals two. However, the assumption of rational roots for these squares leads to a contradiction, and Aristotle quotes this as an example of proof by negation: "The diagonal of the square is incommensurate with the side, because odd numbers are equal to even if it is supposed to be commensurate" [139]. Plato mentions the irrationality of non-square numbers up to the root of 17: "Theodorus here was proving to us something about square roots, namely, that the sides (or roots) of squares representing three square feet and five square feet are not commensurable in length with the line representing one foot; and he went on in this way, taking all the separate cases up to the root of seventeen square feet. There for some reason he stopped" [122]. He goes on to explain that the roots of all numbers which are not squares are irrational.

The discovery of irrational numbers had important consequences. First, the impossibility of representing them by the ratios of integers, or, as we would say, by "rational" fractions, led to a search for approximations. It was found that the irrational number has rational neighbours on either side of it: that is to say, we can find two rational numbers, one larger and the other smaller than the irrational, which are of nearly the same value as it and which confine it within narrow limits. This confinement can be infinitely narrowed down so that the irrational number is, as it were, "caught" by rational numbers. The method of approximation replaced the lost Pythagorean harmony and so became a powerful tool for the comprehension of both mathematical and physical realities. Here science learned the great lesson that this reality can only be approached gradually by innumerable approximations.

The discovery of irrational numbers thus led to the rejection of the Pythagorean picture of physical points strung out in a row. It was replaced by the more penetrating concept of the continuum. Every line is infinitely divisible, i.e. the number of points in it is infinite. The problem of infinity—not the infinity of extension, but that of division—opened up a new world to science. The problem was presented in a most provocative form by Zeno, one of the disciples of Parmenides of Elea, in his famous paradoxes of infinity. Some of these paradoxes were solved by Aristotle; but the notion of the continuum was more profoundly understood by the Stoics in the third century B.C. It forms the subject of a later chapter.

As a matter of fact, it was in geometry that Greek mathematics first made use of the method of confinement. A beginning was made by Eudoxus of Cnidus (c. 409-356 B.C.) with his method of exhaustion. In this he was followed by Euclid and Archimedes. The classic example is Archimedes' determination of π by confining the circle between two infinite series of circumscribed and inscribed polygons: the circumference of the circle is the limit common to the perimeters of these polygons. The application of this method is inseparably bound up with the concepts of continuity and infinity. But it was not till modern times that the difficulties involved in this dynamic conception of reality were finally overcome. Greek science, with its essentially static approach to reality, with a few exceptions only touched the surface of the problem, as we shall see later.

We shall now return to the Pythagorean doctrine of proportions, which is connected with the discoveries they made in the field of musical harmony. This is the first instance of the application of mathematics to a basic physical phenomenon. Its results were regarded by Pythagoreans as decisive confirmation of their teaching about Number as the basis of reality. It is their justifiable pride in these successes that explains their horror of the irrational numbers. The laws of musical harmony were deduced from a series of experiments which are especially noteworthy in view of the paucity of systematic experimentation throughout the Ancient Greek period. We owe our information on this point mainly to the Pythagorean Archytas, a native of Tarentum and friend of Plato, as reported by Porphyry in the third century A.D. Another source is the writings of Theon of Smyrna, who lived in the second century A.D. Theon reports that Pythagoras and his disciples carried out experiments with strings of various length and thickness, and that they also varied the tension of the strings by turns of the screws to which they were attached. Experiments were likewise made with wind instruments of various lengths, and with vessels identical in shape which were filled with different volumes of water, thus producing vibrations of columns of air varying in length. Some of these experiments were merely qualitative. But those with strings and wind instruments of different length were genuine quantitative measurements. The chief result was the discovery of three consonances: the octave, the fifth and the fourth. The ratios of length found for the octave were 1 : 2, for the fifth 2 : 3; and for the fourth 3 : 4. The highest note is produced by the shortest string, and the lowest by the longest. The Pythagoreans were also quite clear about the relation between pitch and frequency, i.e. the rate of vibration of the strings: "In the school of Eudoxus and Archytas it was taught that the law of musical harmony depends on numbers. They also taught that the ratios depend on the movements (of the strings): a swift movement produces a high note, because it vibrates continuously and impinges on the air with greater frequency; whereas a slow movement produces a low note, because it is more sluggish" [25]. This shows the high number was associated with the high note, viz. the short string, in accordance with the law that the frequencies are in inverse proportion to the length of the strings. This is confirmed by a sentence from Archytas: "If one takes a rod and strikes an object slowly and feebly, he will produce a low note with the blow, but if he strikes quickly and powerfully, a high note" [21]. We may notice in passing that other parts of this same passage from Archytas show that the Pythagoreans still held inaccurate views about the propagation of sound in air: they mistakenly thought that the speed of propagation depended on the pitch of the note. But here we are chiefly concerned with the bearing of musical consonances upon Pythagorean mathematics.

The ratios of the consonances mentioned above are 1 : 2, 2 : 3,

3 : 4. They are composed of the numbers 1, 2, 3, 4 which make up the Pythagorean "quartet" (tetractys) and constitute the first four rows of the triangle of points, adding up to ten. The number ten was assigned a place of particular honour in the Pythagorean creed of numbers and will claim our attention further in connection with the Pythagorean cosmology. From the numbers of the quartet one can also construct two fundamental mathematical proportions. The numbers 1, 2, 3 stand in arithmetical proportion to each other, viz. the difference between the first and second equals that between the second and third. Between the numbers 1, 2, 4 there is a geometrical proportion, viz. the ratio of the first to the second equals that of the second to the third. Moreover, the third fundamental proportion, the harmonic, is contained in the three numbers 3, 4, 6, which form the basis of the octave (3 : 6), the fifth (4 : 6) and the fourth (3 : 4). The harmonic proportion is defined as follows: the difference between the second number and the first is to the first as the difference between the third and the second is to the third.

The precise definition of these proportions is given in a passage from Archytas, where their connection with music is specifically stressed. It is worth mentioning that they can also be represented as means: the arithmetical mean of a and c is $b=\frac{1}{2}(a+c)$, the geometrical mean is $b=\sqrt{ac}$, while the harmonic mean is defined by the equation $\frac{1}{b}=\frac{1}{2}\left(\frac{1}{a}+\frac{1}{c}\right)$. Iamblichus (c. 300 A.D.) reports:

"Formerly, in the time of Pythagoras and his mathematical school, only three means were known—the arithmetical, the geometrical and a third which was first called 'subcontrary'. This last was subsequently renamed 'harmonic' by the school of Archytas and Hippasus because it embraces harmonic ratios" [26]. For the sake of completeness it should be added that the Pythagoreans also discovered the harmonic mean in the dimensions of the cube: "There are those who say that it should be called a harmonic mean, after Philolaus, because it inheres in the whole of geometrical harmony. This geometrical harmony, according to them, is the cube which is perfectly harmonious and equal in its three dimensions. This mean reflects itself in the whole cube: the number of sides in the cube is 12, its corners are 8 and its faces 6. Now 8 is the harmonic mean of 6 and 12" [27].

Indeed, the ratio 6 : 8 : 12 equals that of 3 : 4 : 6 which make up the basic consonances.

From these few examples it will easily be understood how the theory of numbers, in the school of Pythagoras, developed into a whole philosophy in which scientific elements in the form of mathematical and physical discoveries were combined with a religious awareness of the unity of the cosmos as expressed in numerical harmony. The following fragments of Archytas make it clear that the principle of number won recognition in much the same way as any modern scientific theory: a general law is first inferred from single facts and then leads to the discovery of additional details and interrelations which in their turn serve to confirm the law: "Mathematicians seem to me to have excellent discernment, and it is in no way strange that they should think correctly concerning the nature of particular existences. For since they have passed an excellent judgment on the nature of the Whole, they were bound to have an excellent view of separate things. Indeed, they have handed on to us a clear judgment on the speed of the constellations and their rising and setting, as well as on geometry and Numbers and solid geometry, and not least on music; for these mathematical studies appear to be related. For they are concerned with things that are related, namely the two primary forms of Being" [21]. Archytas almost certainly referred here to number and measure by which the universe is controlled.

By a quite natural development, the universality of number led the Pythagoreans to project the findings of their theory of musical harmony on to the heavens. For them the cosmos was an ordered system which could be expressed in numerical ratios and which had partly revealed itself in the connection between the lengths of vibrating strings and their notes. The planets revolve in circles in the sky, at varying distances from the centre and at different speeds. This forces on the mind an analogy with music: the movements of the planets in their courses could be compared to the vibrations of the strings, and their angular velocities—to the frequencies of these vibrations. If this analogy has any meaning there should be harmonic ratios in the dimensions of the heavens too, analogous to the ratios of pure consonances. Thus was born the idea of the music of the spheres. "They said of the bodies which revolve around the centre that their distances from it observe certain ratios. Some revolve faster and some slower, the slower giving out a low note in their movements and the faster a high note. Hence, these notes, which depend on the proportions of the distances, together make up a harmony. . . . If the distance of the sun from the earth, for example, is twice that of the moon, and that of Venus three times as great, and that of Mercury four times, they infer that there is a certain arithmetical proportion for the other planets as well, and that the movement of the heavens is harmonious" [28].

Scientific daring, poetical depth and religious fervour combined to provide this theory with such a powerful appeal that it continued to fascinate thinkers right up to modern times. Kepler gave it new life by working the harmony of the spheres into his conjectures about the connection between the planetary orbits and the Platonic bodies. Unlike the Pythagoreans, however, he did not accept the music of the spheres in its literal sense, thus circumventing the question which had been asked of the Pythagorean School: why do we not hear the sounds of the stars in their courses? The Pythagoreans gave an extremely adroit answer, as we learn from Aristotle: "These results clear up another point, namely that the theory that music is produced by their movements, because the sounds they make are harmonious, although ingeniously and brilliantly formulated by its authors, does not contain the truth. It seems to some thinkers that bodies so great must inevitably produce a sound by their movements: even bodies on the earth do so, although they are neither so great in bulk nor moving at so high a speed, and as for the sun and the moon, and the stars, so many in number and enormous in size, all moving at a tremendous speed, it is incredible that they should fail to produce a noise of surpassing loudness. Taking this as their hypothesis, and also that the speeds of the stars, judged by their distances, are in the ratios of the musical consonances, they affirm that the sound of the stars as they revolve is concordant. To meet the difficulty that none of us is aware of this sound, they account for it by saying that the sound is with us right from birth and has thus no contrasting silence to show it up; for voice and silence are perceived by contrast with each other, and so all mankind is undergoing an experience like that of a coppersmith, who becomes by long habit indifferent to the din around him" [29]. No attempt was made by the school of Pythagoras to support the theory of planetary harmony by empirical facts. Nor could its opponents disprove it numerically. But the idea itself and its great popularity are of paramount importance for our understanding of the cosmos as conceived by the Greeks. This belief in the unity of geometry, music and astronomy is an expression of the further belief in the unity of man and the cosmos. There is close connection between the laws of musical harmony, built as they are on simple geometrical or numerical proportions, and such a fundamental physiological and psychological fact as the sense of euphony. "Music is the arithmetic of the soul, which counts without its being aware of it"—as Leibniz put it. This is a question which we are still far from fully understanding, in spite of Helmholtz's theory of musical harmony. What, for example, is the secret of the harmony of the octave in which we feel the identity of the note repeated on a higher level? Why does this peculiarly perfect harmony take the form of the simplest ratio of all, 1 : 2? Would a similar sensation be aroused optically, if the visible spectrum covered more than an octave?

Though the Greeks were probably unaware of all these problems, they intuitively sensed that here they had a glimpse of the bonds that connect human consciousness and the outside world, some indication of man's participation in the scheme of the cosmos. Modern science, while not denying the reality of such participation, regards it differently. To-day we aim at projecting the mathematical and physical laws of the physical universe into man, with the object of explaining the phenomena of life by physics and mathematics; whereas the Greeks sought to extrapolate man into the expanse of the cosmos and regarded the cosmos as a living organism. Their biological metaphors, such as the breathing of the cosmos, are not simply allegorical: they really mean that the cosmos has its own rhythm of life, that its laws are basically organic and that therefore it is conscious of the musical harmony of the spheres. The conception of the world as a living body was present in all periods of Greek science. Any deviating tendency, such as the atomic theory, did not take firm root in the science of the Ancient World. We shall see later that herein lies the essential difference between it and modern science, despite the similarity of their formal scientific approach.

This similarity cannot just be disregarded. If we wish to make the picture complete, we must study the points of resemblance no less then the points of difference. The first mathematico-physical law discovered by the Greeks was a law of pure numerical proportions, as it treats of the relations between qualities of the same dimension, such as the ratios of the lengths of strings or those of musical frequencies. A second law also expressing a mathematical proportion, which was formulated almost three hundred years later, was Archimedes' law of the lever. This law, too, deals with pure numerical ratios, since it compared the lengths of the arms of the lever and the weights acting upon them with each other. Modern physics, on the contrary, started with the definition of dimensional quantities such as velocity, force, pressure, etc., which are built up of basic dimensions as e.g. length, time or mass. Although these quantities have not the same general significance as pure numbers, they are of inestimable importance as tools for the comprehension of nature. Actually, the Greeks described only two such dimensional quantities: Aristotle recognized the importance of velocity, and Archimedes defined specific gravity—the ratio of the weight of a body and its volume. Before Archimedes, Greek physics had for nearly four hundred years been unable to free itself, despite all efforts, from the vicious circle of "heavy" and "light", because of its inability to define specific weight. But even the discoveries made by Archimedes through this new concept could not entirely do away with the absolute antithesis of heavy and light. The same was true of the notion of velocity which remained extremely primitive. Yet no better proof of the Pythagorean idea that number underlies physical attributes could be found than the physical quantities which are expressed by dimensions. For a body is differentiated simply and solely by the sum total of these quantities which define its physical attributes: specific weight, specific heat, the constants of elasticity, viscosity, etc. In this respect the full implications of Pythagoras' conception were not realized until modern times. But we have gone even further and discovered that certain quantities of this kind have a significance which goes far beyond their meaning for the particular phenomenon in which they were discovered. Such quantities are called "universal constants", regardless of the fact that they have a particular dimension. An example of such a universal quantity is the velocity of light which appears in mechanics, optics, electricity and atomic theory. Another of the very few quantities belonging to this class is Planck's quantum of action. Of even greater importance is the still smaller number of non-dimensional quantities, especially when they are of universal significance. These, as we have seen, are pure numbers, i.e. they appear as ratios of quantities of the same dimension. For example: the product of the velocity of light and Planck's constant has the same dimension as the square of the charge of the electron whose importance as a "quantum of electricity" is unquestioned. Hence the ratio of these quantities is a pure number (with the approximate value of $\frac{1}{137}$). No one doubts to-day that this number is of intrinsic, if still imperfectly understood, significance for the structure of the physical world. Some such feeling of having lighted upon a key to a deeper understanding of nature must have been experienced by the Pythagoreans when they discovered the ratios of musical harmony. We, like them, proceed on the assumption that physical laws discovered in the laboratory, through man-made instruments, are of the nature of cosmic laws. And in spite of the present speculative state of modern cosmology, we are inclined to attach importance to certain numerical ratios between atomic and cosmic quantities, although we have not yet succeeded in explaining them satisfactorily.

We may say, then, that, from the standpoint of modern science, the scientific method of the Pythagoreans was correct. True, the special field of investigation chosen by them does not appear to us to be a suitable starting-point for a systematic study of the physical world as a whole. Yet we observe that their approach was the same as ours: starting from experiments on various instruments, they expressed their findings in general terms and aimed at the mathematical formulation of universally applicable laws. Their fondness for numerical mysticism in no way detracts from their scientific qualities; on the contrary, in so far as it stimulated them to search for causal laws in nature, it contributed to the progress of science. The theoretical physicists of our time, who have so greatly advanced the understanding of nature, likewise believe almost religiously in the power of the mathematical symbol and the validity of formulae and calculations. For this reason we are so surprised that the Pythagorean beginnings should just have petered out, instead of providing the starting-point for a detailed quantitative study of natural phenomena in all the fields of science, and not only in astronomy. What was the cause of the decline of the Pythagorean School and the "religion of numbers"? And why did the mystic element in it eventually prevail over the scientific?

There is no easy answer to these questions. But it can hardly be doubted that a decisive factor here was the enormous influence of Plato, whose philosophy was not such as to encourage any further development along the scientific lines of the Pythagorean School. This is a fascinating question, since there are conspicuous Pythagorean elements in Plato's philosophy, as in the repeated emphasis in the dialogues on the importance of number, arithmetic and measurement. In the *Philebus*, for example, Socrates asks: "And in the productive or handicraft arts, is not one part more akin to knowledge, and the other less; and may not the one part be regarded as the pure, and the other as the impure?" And he continues: "I mean to say, that if arithmetic, mensuration, and weighing be taken away from any art, that which remains will not be much" [124]. Nor can Plato be charged with turning his back on nature: not a few passages in various dialogues, where he employs undeniably physical metaphors, prove just the opposite. It would be as well to illustrate this by two such passages, in one of which he describes magnetic attraction and in the other capillary forces: "It is a divine influence which moves you, like that which resides in the stone called magnet by Euripides, and Heraclea by the people. For not only does this stone possess the power of attracting iron rings, but can communicate to them the power of attracting other rings; so that you may see sometimes a long chain of rings, and other iron substances, attached and suspended one to the other by this influence" [127]. And capillarity is described as follows: "How fine it would be . . . if wisdom were a sort of thing that could flow out of the one of us who is fuller into him who is emptier, by our mere contact with each other, as water will flow through wool from the fuller cup into the emptier" [125].

These examples are evidence of a closeness to nature and of an interest in at least qualitative observations. Yet, for all that, Plato's basic conception that truth is to be found only in the world of pure forms, in the realm of the "ideas", predisposed him to believe that number and the study of number also have no value as long as the number inheres in physical substance, in the world of shadows. Further on in the conversation quoted from the *Philebus*, after it has been shown that the scientific element in the arts is essentially the mathematics in them, Socrates persuades his friend that there are two kinds of arithmetic. "What is the science of calculation or measurement used in building or commerce, when we compare them with the philosophy in geometry and exact reckoning?" Finally, after he has reiterated that the arts belonging to mathematics are far superior to all others, he concludes: "The arts and sciences which are animated by the true philosophy are infinitely superior in accuracy and truth to measures and numbers" [124]. The complete separation of the world of the ideas from our physical world involved, in Plato's system of thought, a distinction between the Pythagorean numbers which inhere in concrete objects and number as such. The Pythagorean numbers too are impure, so to speak; and there is no sense in physical measurements, seeing that there is no connection between them and the absolute precision of the pure number. This philosophy led Plato, and all the subsequent generations educated on his doctrine, to treat quantitative experimentation with contempt and to give up the aim of expressing physical facts in terms of number. Thus the original position of Pythagoras was undermined and his specific demand for the mathematical description of reality was only very imperfectly fulfilled throughout ancient times. Nowhere does Plato express his opinion more clearly and vehemently than in the seventh book of the *Republic*: "Anyone can see that this subject [astronomy] forces the mind to look upwards, away from this world of ours to higher things" (says Glaucon). "Anyone except me" (says Socrates). "I cannot think of any study as making the mind look upwards, except one which has to do with unseen reality. No one, I should say, can ever gain knowledge of any sensible object by gaping upwards any more than by shutting his mouth and searching for it on the ground, because there can be no knowledge of sensible things. . . . These intricate traceries in the sky are, no doubt, the loveliest and most perfect of material things, but still part of the visible world, and therefore they fall far short of the true realities—the real relative velocities, in the world of pure number and all perfect geometrical figures, of the movements which carry round the bodies involved in them. These, you will agree, can be conceived by reason and thought, not seen by the eye. . . . Accordingly, we must use the embroidered heaven

44

as a model to illustrate or study those realities. . . . If we mean, then, to turn the soul's native intelligence to its proper use by a genuine study of astronomy, we shall proceed, as we do in geometry, by means of problems, and leave the starry heaven alone. . . ." After this, Socrates turns against the Pythagoreans: "As you will know, the students of harmony make the same sort of mistake as the astronomers: they waste their time in measuring audible concords and sounds one against another. . . . You are thinking of those worthy musicians who tease and torture the strings, racking them on the pegs. . . . They are just like the astronomers—intent upon the numerical properties embodied in these audible consonances: they do not rise to the level of formulating problems and inquiring which numbers are inherently consonant and which are not, and for what reasons" [128].

In his sweeping irony at the expense of the Pythagorean scientific method, Plato ignores the fact that, in science, the inner sight is not in itself sufficient for the discovery of truth. He forgets that only those very experiments of which he makes fun made it possible for the Pythagoreans to formulate their mathematical laws. The theory of ideas, however, did not regard experiment as a means to the desired goal. The great popularity of this philosophy of Plato's, which so deplorably helped to delay the synthesis of the experimental method with mathematics, is also to be largely explained by the Greek tendency to overestimate the power of deduction to such an extent that induction seemed to become wholly unnecessary.

Of the very few exceptions to this rule, the greatest was Archimedes (287-212 B.C.). Amongst his mathematico-physical discoveries, the laws of hydrostatics are perhaps more important than the laws of the lever: the former include the use of a dimensional physical quantity, viz. density, whereas the latter are no more than laws of pure proportion, like the law of vibrating strings. In another respect no progress was made in the period from Pythagoras to Archimedes: the natural laws which both of them formulated are static laws in which time does not appear explicitly. Indeed, apart from the movements of the stars, not a single dynamic phenomenon was put into mathematical form by the Greeks. Aristotle, in formulating a general theory of dynamics, discusses the laws of motion in a mainly qualitative way, although he was aware of the numerical nature of time, as we see in the fourth part of his *Physics*. He recognized that there is connection between time and motion: "We must take this as our starting-point and try to discover—since we wish to know what time is—what exactly it has to do with movement. Now we perceive movement and time together: for even if it is dark and we are not being affected through the body, if any movement takes place in the mind we at once suppose that some time has also elapsed; and not only that but also, when some time is thought to have passed, some movement also along with it seems to have taken place. Hence time is either movement or something that belongs to movement. Since then it is not movement, it must be the other" [146]. After discussing the continuity of motion and the concepts "now", "before", "after", he comes to the following conclusion: "Time is just this—number of motion in respect of 'before' and 'after'. Hence time is not movement, but only movement in so far as it admits of enumeration. A proof of this: we discriminate the more or the less by number, but more or less movement by time. Time then is a kind of number" [147]. Every measurement of time, whether by the observation of the stars or by a man-made watch, depends on motion, and Aristotle stresses the interrelation of the concepts: "Not only do we measure the movement by the time, but also the time by the movement, because they define each other. The time marks the movement, since it is its number, and the movement the time" [148].

For all that, Aristotle never managed to formulate his ideas mathematically as quantitative laws of motion. He did not fully understand Pythagoras' method of explaining the sensory world in mathematical terms; and he would almost certainly not have accepted the hydrostatics of Archimedes. He criticizes the disciples of Pythagoras for being unable to explain motion by mathematics, but his criticism shows that he had missed the real point of the Pythagorean method. Unlike his master Plato, Aristotle attached importance to the sensory world; but like Plato he did not see what quantitative experiment and mathematical formulation had to do with sensory objects: "The Pythagoreans treat of principles and elements stranger than those of the physical philosophers (the reason is that they got the principles from non-sensible things, for the objects of mathematics, except those of

46

astronomy, are of the class of things without movement); yet their discussions and investigations are all about nature; for they generate the heavens, and with regard to their parts and attributes and functions they observe the phenomena, and use up the principles and the causes in explaining these, which implies that they agree with the others, the physical philosophers, that the real is just all that which is perceptible and contained by the so-called 'heavens'. But the causes and the principles which they mention are, as we said, sufficient to act as steps even up to the higher realms of reality, and are more suited to these than to theories about nature. They do not tell us at all, however, how there can be movement if limit and unlimited and odd and even are the only things assumed, or how without movement and change there can be generation and destruction, or the bodies that move through the heavens can do what they do. Further, if one either granted them that spatial magnitude consists of these elements, or this were proved, still how would some bodies be light and others have weight? To judge from what they assume and maintain they are speaking no more of mathematical bodies than of perceptible; hence they have said nothing whatever about fire or earth or the other bodies of this sort, I suppose because they have nothing to say which applies peculiarly to perceptible things" [30].

These sentences show that, even if the disciples of Pythagoras had succeeded in giving a quantitative formulation to dynamic phenomena, Aristotle would not have accepted this as a solution of the problem. He wanted to explain literally "how there can be movement" and not "how such and such a motion is dependent on such and such factors" or "what form the motion takes in such and such circumstances". Questions of this kind can be solved by Pythagoras' method; whereas Aristotle's attitude leads nowhere and offers no hope of fruitful research in the natural sciences. Moreover, the generalizing tendency, which had so greatly assisted early scientific thought in Greece, was carried so far by Aristotle that it became a hindrance. He started from a generalization without first having submitted to careful quantitative examination many details on which the generalization should be based. His dynamics, a characteristic instance of this weakness, will be dealt with in another chapter. But this alone is not sufficient to explain why the full scope of the connection between mathematics and sensory phenomena eluded the Greeks.

The relation between mathematics and experience also engaged the attention of Aristotle's friend and pupil Theophrastus, who states the problem very lucidly at the beginning of his essay on metaphysics: "How and in what terms should we define the philosophy of the first things? For the science of nature is more complex and, as some hold, less orderly as being subject to all manner of changes. Whereas the philosophy of the first things is clearly defined and unchanging. Hence it is associated with the mind, and not with the senses, since the mind is immovable and unchangeable. In general it is considered greater and loftier than natural science. The first question, then, is whether there is a connection and a kind of interaction between mental concepts and natural phenomena; or whether there is none, but somehow both, while remaining separate, work together towards the elucidation of reality. It is obviously more rational to assume that there is some connection and that the universe is not incoherent, but has first and last, essential and less essential, just as, for example, what is eternal takes precedence over what is transitory. If so, what is the nature of these things and in what does it consist? For if mental concepts belong solely to mathematics, as some say, the connection with the senses is not very clear nor does it seem capable of actually performing anything. For mathematics would seem to be constructed by us who put figures, shapes and ratios into things which in themselves do not exist in nature and therefore cannot unite with natural objects so as to produce life and movement in them. Even number itself cannot do this, even though it is held by some to be the first and governing principle" [174].

In these few lines Theophrastus gave the classic formulation of a whole complex of problems. It can be seen that, towards the end, he too, like his master Aristotle, is inclined to doubt whether mathematics can explain movement. Much more penetrating is his general question whether natural phenomena can be expressed in terms of the artificial language of mathematics: how can what is natural be explained by what is not natural, like by unlike? In raising this epistemological question Theophrastus entered on the border-land between Ancient Greek and modern science. He believed, as we see from the opening sentences, that there is an interaction between mind and the sensory world. He further believed that the world "is not incoherent", i.e. that it is a well-ordered cosmos. But he is baffled by the contrast between the simplicity and orderliness of the mind and the complexity and ramification found in most natural phenomena. How does the mind master them all? To-day we might perhaps set Theophrastus' mind at rest by answering him with the famous metaphor used by his first teacher, Plato. In the *Timaeus* Plato describes how the demiurge forms the cosmos out of unformed primeval matter. The creator here symbolizes the Mind, while the raw material mastered by the mind is given by Plato a quality which he calls "necessity". This has nothing to do with the concept of necessity in Greek philosophy, where it is entirely concerned with causality. "Necessity" in the *Timaeus* denotes the resistance of the primeval matter to order and form: "Now our foregoing discourse . . . has set forth the works wrought by the craftsmanship of Reason; but we must now set beside them the things that come about of Necessity. For the generation of this universe was a mixed result of the combination of Necessity and Reason. Reason overrules Necessity by persuading her to guide the greatest part of the things that become towards what is best; in that way and on that principle this universe was fashioned in the beginning by the victory of reasonable persuasion over Necessity" [131]. This metaphor aptly describes the mathematical comprehension of nature as we see it, for example, in the history of mathematical physics over the past three hundred years. This process of interaction goes on infinitely; Reason and Necessity always exist side by side, no matter how much the former expands. This expansion is achieved by enticement, so to speak: Necessity has to be gradually persuaded into obeying Reason. Although the mathematical description of the universe is scientific in function, its attainment is impossible without the inspired imagination of the creative scientist. Step by step the realm of explanation slowly imposes its sway upon the chaos of the unexplained. The task of conquest was begun by mechanics; then thermodynamics reduced another sphere of phenomena to order; the work was continued by electromagnetism; and now new horizons have been opened up by the quantum theory. It is true that, after every conquest, Necessity makes its appearance in another field; but we are not daunted or dismayed. To-day we no longer doubt that it is possible "to guide the greatest part of things that become towards what is best".

III

HEAVEN AND EARTH

"Knowest thou the ordinances of the Heavens? Canst thou
establish the dominion thereof in the earth?"   JOB 38.33

THE Greek contribution to astronomy, the most ancient of
all the sciences, is particularly marked in the following three
spheres: (*a*) the improvement of astronomic measurements; (*b*)
the development of geometrical models for the explanation of
stellar movements; (*c*) the calculation of cosmic dimensions. In
the first case the Greeks merely carried on from the point reached
by the Egyptians and Babylonians. But in the other two they
opened new chapters in the history of astronomy which resulted
in far-reaching advances.

The best criterion of the improvement made in astronomical
observations during the Greek period is the degree of accuracy
attained in determining the length of the year. The solar year
of 365 days was known to the Greeks at the end of the sixth
century B.C. from Egyptian sources, just as they got their knowledge of the monthly period from Babylonia. (From the earliest
times the Babylonians had studied the connection between the
lunar and solar cycles.) The Greek astronomers of the fifth
century B.C. narrowed down the earlier approximation by
observations of the solstices on the shortest and longest days of
the year. In 432 B.C. Meton of Athens estimated the length of the
year as 365 days, 6 hours, 18 minutes, 56 seconds, a figure which
is only 30 minutes, 10 seconds more than the correct value. In
the next three hundred years the margin of error was steadily

50

reduced, until Hipparchus, whose main observations were completed by about 130 B.C., arrived at the figure of 365 days, 5 hours, 55 minutes, 12 seconds—an error of only 6 minutes, 26 seconds. Thus the accuracy was of the order of approximately one part in one hundred thousand, and still more accurate was the mean length of the month as known in the time of Hipparchus. In Meton's cycle 19 solar years (including 7 leap-years) were equal to 235 lunar months, while in Hipparchus' cycle there were 304 years (including 112 leap-years) which equalled 3760 lunar months. This degree of accuracy, which so remarkably demonstrates the patient perseverance of the Greek astronomers in their observations of the heavens, set the standard for mediaeval and modern workers in the same field.

Bearing in mind the paucity of precise measurements made by the Ancient Greeks in all the other physical sciences, we cannot help asking why astronomy was an exception. It is usual to stress the practical, and particularly the economic, factors which led to the development of astronomical research. The importance of the stars to navigation is frequently mentioned in ancient poetry; and there is literary evidence from very early times for the connection between agriculture and the knowledge of the heavens. Hesiod, for instance, advises the farmers on the periods of sowing, reaping and grape-picking by the risings of certain constellations after sunset or before sunrise. Similarly, the Egyptian peasant had from time immemorial learnt to calculate the beginning of the Nile's flood by the stars. But more important than these practical considerations were irrational factors the roots of which go down to still earlier times and which are connected with the history of astrology. Especially noteworthy in this connection is the effect produced on man by the cyclical character of heavenly phenomena. The changes in the phases of the moon and their periodicity, the progress of the sun along the zodiacal belt with its attendant changes of season, the complicated movements of the planets which also have a cyclical regularity, and above all the twenty-four hours periodicity of the whole heavenly dome with the accompanying alternations of day and night—all this vast array of eternally recurring cycles awoke the consciousness of ancient man to the great contrast between the firm-set certainty of the heavens and the uncertainty of human life on the earth.

"Canst thou lead forth the Mazzaroth in their season?
Or canst thou guide the bear with her train?
Knowest thou the ordinances of the heavens?
Canst thou establish the dominion thereof in the earth?"

(Job 38.32-33.)

Of the many examples from classical literature we shall content ourselves with the verses from one of Horace's most beautiful odes (IV.7) in A. E. Housman's translation:

"But oh, whate'er the sky-led seasons mar,
Moon upon moon rebuilds it with her beams:
Come *we* where Tullus and where Ancus are
And good Aeneas, we are dust and dreams."

Here there is a twofold contrast: man's life is unique and can be lived only once, whereas in the heavens there is endless recurrence; the individual is the plaything of chance, whereas the stars obey a law of marvellous constancy. This contrast is the source of that religious fascination which from the earliest times held man's mind in thrall to the heavenly bodies, expressing itself as simple star worship or in a rationalized form as an urge to follow the minute details of those manifestations of precision and regularity. To this irrational element there was added the practical aim of fixing the religious festivals by the calendar. The best known example from modern times is the official support given by the Catholic Church to the astronomical observations which resulted in the introduction of the Gregorian calendar in 1582.

This union of science and religion in the approach to the phenomena of the heavens was disturbed by the new, intensely rational, scientific method introduced by the Milesian School in the very early days of Greek science. We have seen that Anaximander explained the revolutions of the sun and the moon with the aid of mechanical models; moreover, Anaxagoras said that the sun and the stars were flaming stones, and similar opinions are attributed to Leucippus and Democritus, the founders of the atomic school. Some of Empedocles' conjectures about heavenly phenomena are also purely physical in character. Still, even at that time, the religious notion was not entirely excluded from scientific thought; indeed, the scientific sublimation of star-worship appears still more strongly in the Pythagorean School, where religious and mystic tendencies prevailed and combined with the scientific approach. Of the Pythagorean physician Alcmaeon (beginning of the fifth century B.C.), Aristotle says in the course of his discussion of the nature of the soul: "Alcmaeon also seems to have held a similar view about the soul; he says that it is immortal because it resembles 'the immortals', and that this immortality belongs to it in virtue of its ceaseless movement; for all the 'things divine', moon, sun, and the planets, and the whole heavens, are in perpetual movement" [31]. The cyclical motion of the heavenly bodies displays this combination of immortality and continuity in its purest form, thus giving dynamic proof of the divinity of the stars.

This linking together of the soul and the stars which is found also in a late source ("Alcmaeon of Croton regarded the stars as gods, since they have souls" [32]), had a characteristic sequel in Plato's teaching. His famous proof of the immortality of the soul begins with the following sentence: "Every soul is immortal—for whatever is in perpetual motion is immortal" [126]; and his view about the divinity of the stars found its classic expression in the *Epinomis*. In this culmination of the scientific sublimation of star-worship along Pythagorean lines we may also see some elements of the foundation on which Aristotle's dynamics was built: "Men ought to have regarded the possession of intelligence by the stars and all their movements as proved by the uniformity of their action, and by the fact that they continue to carry out the counsels formed long ago, and do not wander about with varying revolutions, altering their counsels this way and that, and doing first one thing and then another. The majority of us have adopted exactly the opposite view, thinking, because of the uniformity of their actions, that these beings have no soul; the multitude thus follows the lead of fools, and considers that the human race shows intelligence and life by its mutability, whereas that of the gods is devoid of intelligence, because it remains in the same orbits. Man might, however, have adopted a view nobler, better, and acceptable—the view that that which acts always in the same uniform way and under the influence of the same causes, should on this very account be regarded as possessed of intelligence; and this specially applies to the nature of the stars, which form so glorious a spectacle, and, while performing the movements of their dance, of all dances the most lovely and magnificent, discharge their duty to all things that live. And, as a proof that we are right in ascribing life to them, let us first consider their size. For they are not small, as they appear to the eye, but each of them is immense in bulk; this we are bound to believe, for it is established by sufficient demonstrations. The sun as a whole may be rightly regarded as larger than the earth as a whole; and all the stars in their courses are of wonderful size. Let us consider how it could be that anything could cause so great a bulk to move in its orbit in exactly the same time in which it now performs its course" [136].

We of the machine age have grown accustomed to an entirely different association of ideas. The essence of every machine is that it repeats the same movement exactly; so that we use the expression "automatic" to indicate precisely a movement that is devoid of reason, a "soulless" movement. But in the age of arts and handicrafts, the exact reproduction of a model or form was regarded as a sign of the artist's divine inspiration. It was Plato's educational influence that decided in favour of the view that the stars are divine and against the purely physical tendency of the pre-Socratic period. Plato himself explicitly argues against the opinions of Anaxagoras in the tenth book of the *Laws*: "When you and I try to prove the existence of the gods by pointing to these very objects—sun, moon, stars, and earth—as instances of deity and divinity, people who have been converted by these scientists will assert that these things are simply earth and stone, incapable of paying any heed to human affairs" [132].

In the twelfth book Plato sums up his views about the existence of the gods. He finds two proofs for this: the immortality of the soul which results from its perpetual motion, and the movement of the heavens. He asks if it is conceivable that ". . . those who study these objects in astronomy and the other necessary allied arts become atheists through observing, as they suppose, that all things come into being by necessary forces and not by the mental energy of the will aiming at the fulfilment of good" [135]. His answer is that: "The position at present is exactly the opposite of what it was when those who considered these objects considered them to be soulless. Yet even then they were objects of admiration, and the conviction which is now actually held was suspected by all who studied them accurately—namely, that if they were soulless, and consequently devoid of reason, they could never have employed with such precision calculations so marvellous" [135].

This polemic, too, is directed against Anaxagoras and the physical way of looking at the heavens, as is shown by the next passage, in which it is reiterated that "all that moves in the heavens appeared to them to be full of stones, earth and many other soulless bodies which dispense the causes of the whole cosmos". At the end there appears the Pythagorean element in Plato's philosophy: "And in addition to this, as we have often affirmed, he must also grasp the reason which controls what exists among the stars . . . and he must observe also the connection therewith of musical theory . . . and he must be able to give a rational explanation of all that admits of rational explanation" [135].

We have quoted at length from Plato, because it was his opinion that decided this important dispute. Once Aristotle had accepted it and given it a broader physical basis, the fate of Greek science was sealed and the division between heaven and earth became an integral part of ancient physics and the Greek cosmos. This position remained unchanged till Galileo. Even Epicurus, who tried to follow in the steps of Anaxagoras and Democritus and therefore contested Plato's views about the divinity of the stars, maintained this separation, though in exactly the opposite sense. The accepted view was that the soul and intelligence of the stars was revealed in the conformity to law and absolute regularity of their movements. Epicurus therefore tried to undermine the Platonic theory by directly and indirectly raising doubts about the existence of such conformity to law in the phenomena of the heavens, while maintaining that terrestrial phenomena were strictly obeying the causal laws. Modern science was born in the seventeenth century, when it was shown that the laws of terrestrial mechanics hold good also for the movements of the planets; in other words, with the breaking down of the barrier which separated heaven from earth. That is why the history of this separation is so important for the understanding of the cosmos as pictured by the Ancient Greeks. We shall have to return to it several times, especially in connection with Aristotle, who gave this separation a general scientific sanction and whose great intellectual authority preserved it for nearly 2,000 years; and also in connection with Epicurus, who kept it in being by going to the opposite extreme.

The highest degree of accuracy in astronomical observations was reached by the greatest of the Greek astronomers, Hipparchus (*c.* 190-120 B.C.). Some of his observations were made in Rhodes and Alexandria and his findings formed the basis of Ptolemy's compendium about three hundred years later. Hipparchus made systematic use of trigonometry in his calculations and for this purpose constructed a "Table of Chords" giving numerical values of various arcs or angles of a circle. Besides this, he improved some of the measuring instruments and also made a catalogue containing more than eight hundred stars whose positions he determined by means of co-ordinates. This list has not come down to us. Indeed, of all his writings there remain only a few fragments quoted in the works of later astronomers. Pliny, in his *Natural History*, praises him in the following terms: "Hipparchus, who can never be lauded too highly . . . discovered a new star which appeared in his time. Because of its movement on the day of its appearance, he began to wonder if the same thing might not occur frequently and if the stars which are considered fixed might not also move. He did something that would be daring even in a god—he counted the stars and constellations for future generations and gave them names. For this purpose he devised instruments whereby he assigned a position and size to every star. As a result, it is easy to distinguish not only whether stars are dying or being born, but also whether they move from their place and whether their light is increasing or decreasing. He left the heavens as a heritage for all who may desire to take possession of them" [238].

Hipparchus' greatest achievement which assures him a place of honour for all time in the history of astronomy was his discovery of the precession of the equinoxes. This can be explained in a few sentences. The axis of the earth is not perpendicular to the plane of its orbit, but inclined at an angle of 66·5°. Therefore the plane of the equator is inclined at an angle of 23·5° to the plane of the orbit. From the daily revolution of the earth upon its axis it is easy to determine the direction of the axis in relation to the stars and also the projection of the earth's equator as a circle on the sky; whereas the annual revolution of the earth appears in the heavens as the movement of the sun along the orbit of the ecliptic in the zodiacal belt. Hence this orbit cuts the projection of the equator at an angle of 23·5° at two points, which are the vernal and autumnal equinoctial points. According to the laws of mechanics, the earth, not being a perfect sphere, behaves like a top: its axis does not maintain its direction in space, but revolves (very slowly) upon an axis vertical to the plane of its

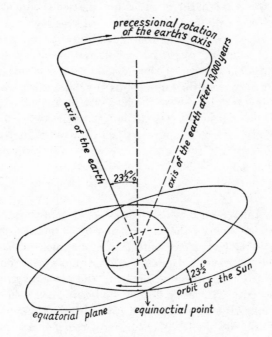

FIG. 2. Precession of the equinoxes.

orbit and passing through its centre. On account of this top-like motion the position of the equinoctial points in the sky changes. This movement is called the precession of the equinoxes. Since the precession completes a full circle in approximately 26,000 years, we have here a movement which involves the annual displacement of a given point in the sky by 50″ of the arc. By this we can estimate the accuracy of Hipparchus' measurements. He arrived at his results by comparing his own measurements of the position of certain fixed stars relatively to the equinoctial points with those of the Alexandrian astronomers during the previous one hundred and fifty years and more. We find that he achieved an accuracy of as much as 15 per cent. in his determination of the precession.

This explains Ptolemy's words about the discovery: "That the sphere of the fixed stars has a movement of its own in a sense opposite to that of the revolution of the whole universe, that is to say, in the direction which is east of the great circle described through the poles of the equator and the zodiac circle, is made clear to us especially by the fact that some stars have not kept the same distance from the solstitial and equinoctial points in earlier times and in our time respectively, but, as time goes on, are found to be continually increasing their distance, measured in the eastward direction, from the same points beyond what it was before. . . . This seems to have been the idea of Hipparchus, to judge by what he says in his work 'On the length of the year': 'If for this reason the solstices and equinoxes had changed their position in the inverse order of the signs, in one year, by not less than $\frac{1}{100}°$ their displacement in 300 years should have been not less than 3°'" [239].

Let us turn now to the great original contribution of the Greeks in astronomy—the construction of geometrical models to depict the planetary movements. Since Copernicus it is common knowledge that these movements appear to us as they do because of our own revolution round the sun. The planets with a larger orbit than ours, viz. Mars, Jupiter and Saturn, appear to move round the earth each on its own course, but this movement is not uniform, and the circular course of the planet is at certain times interrupted by a movement in a loop: the planet retards its movement and turns back, moving for a certain while in the opposite direction; then it stops and once again advances beyond the turning-point, and so on. The planets between us and the sun, viz. Venus and Mercury, also make similar retrograde movements in their apparent revolution round the earth, but it should be noted that the centre of these backward and forward oscillations is the sun (which itself appears to revolve around the earth), and therefore Venus, for example, appears sometimes as the evening-star and sometimes as the morning-star. How can all these movements, including the sun's yearly cycle and the moon's monthly cycle, be reduced to one single system? On top of all these movements there is the daily cycle of the whole sky with all its hosts, due to the daily revolution of the earth upon its own axis. Here the geometrical genius of the Greeks found full scope for its powers. The original inspiration came from the idea of Pythagoras and Plato that the sphere and the circle are the most perfect geometrical figures and therefore the basis of every movement in the heavens.

The first solution to the problem was provided by Eudoxus of Cnidos (c. 409-356 B.C.) under the influence of Plato: "And, as Eudemus related in the second book of his astronomical history, and Sosigenes also who herein drew upon Eudemus, Eudoxus of Cnidos was the first of the Greeks to concern himself with hypotheses of this sort, Plato having, as Sosigenes says, set it as a problem to all earnest students of this subject to find what are the uniform and ordered movements by the assumption of which the phenomena in relation to the movements of the planets can be saved" [115]. "To save the phenomena" is a characteristic Greek expression for the rational explanation of physical phenomena in general, and particularly of astronomical phenomena. Eudoxus' method is the famous theory of concentric spheres which was accepted by Aristotle and through him became part of mediaeval astronomy. The concentric spheres have the earth as their common centre, and they revolve at constant velocities, varying for each sphere, around different axes and in different directions. All these velocities and directions are so co-ordinated that the movement resulting from them is the empirical movement of the planet in question which itself is assumed to be attached to the equator of the innermost sphere. We shall now quote some sentences from the detailed description of Eudoxus' theory given by Aristotle in his *Metaphysics*: "Eudoxus supposed that the motion of the sun or the moon involves, in either case, three spheres, of which the first is the sphere of the fixed stars, and the second moves in the circle which runs along the middle of the zodiac, and the third in the circle which is inclined across the breadth of the zodiac; but the circle in which the moon moves is inclined at a greater angle than that in which the sun moves. And the motion of the planets involves, in each case, four spheres, and of these also the first and second are the same as the first two mentioned above (for the sphere of the fixed stars is that which moves all the other spheres, and that which is placed beneath this and has its movement in the circle which bisects the zodiac is common to all), but the poles of the third sphere of each planet are in the circle which is inclined at an angle to the equator of the third sphere; and the poles of the third sphere are different for each of the other planets, but those of Venus and Mercury are the same" [116]. Thus Eudoxus required 26 spheres to "save the phenomena".

The development of the theory of spheres is in two respects characteristic of the development of all scientific hypotheses. First new facts are accumulated or the accuracy of old information is increased by new observations. Then the necessity becomes felt of adjusting the old model to the new reality. Usually this involves first of all making the model more complex, and eventually abandoning it altogether for another, based on a new idea and simpler (in the scientific sense) than the first. The model of Eudoxus was "improved", i.e. made more complex, by Callippus (370-300 B.C.), who added seven more spheres to the original 26: two each for the sun and moon, and one each for Mercury, Venus and Mars. Here what was gained in accuracy was more than lost in complexity. Of still greater interest is another modification of the theory of spheres. Generally speaking, a model is introduced as an *ad hoc* hypothesis to make possible a first concrete description of the phenomena to be examined. Eudoxus, for example, invented his spheres solely as geometrical pictures which provided a neat solution of the problem posed by Plato. There was apparently no intention of turning this purely geometrical model, which was conceived as a mathematical device, into a picture of physical reality. But sometimes the model is simply a reproduction of reality as pictured by our mind. The planetarium, for example, with its representation of the planets by small balls revolving round the sun, is a miniature model of undoubted physical reality. On the other hand, we are far from being certain about any mechanical model of the particles in the nucleus of the atom, and we regard all such models as no more than *ad hoc* hypotheses, pending further clarification of nuclear problems.

However, models of this kind have the dangerous tendency to acquire a life of their own and become independent. Their original purpose was to assist a theory at a certain stage of its development, but once they entrench themselves in the scientific mentality of their time, they become obstacles to further progress, since they tend to be identified with reality and to be considered even more important than the theoretical foundations of the theory for whose sake they had been invented. This is what happened to Eudoxus' model of the spheres when Aristotle got hold of it and turned the spheres into physical bodies, material concentric spheres joined together. This we learn from the passage in the *Metaphysics* which introduces his description of Eudoxus' theory and from the improvements which he himself made in the theory. He argues that, since the stars are substances, their moving agency must also be a substance, seeing that it is prior to them in time: "That the movers are substances, then, and that one of these is first and another second according to the same order as movements of the stars, is evident. But in the number of movements we reach a problem which must be treated from the standpoint of that one of the mathematical sciences which is most akin to philosophy—viz. of astronomy; for this science speculates about substance which is perceptible but eternal, but the other mathematical sciences, i.e. arithmetic and geometry, treat of no substance" [171].

In this way the theory of concentric spheres was elevated by Aristotle from a mere geometrical aid to the level of a physical reality, an actual reproduction of the cosmos. Thus supported by the authority of the Aristotelian concept of substance and cause, it was destined to maintain an uncontested place in the cosmic picture for eighteen hundred years after Aristotle's death. Like Callippus, Aristotle modified the model of Eudoxus; only, in his case the modifications were not meant to make the model correspond more exactly to observed facts, but to bring it into harmony with Aristotle's theory. Eudoxus had attributed a separate system of spheres to every planet, but had not joined the different spheres together in any way. He had not required any such connection, since the only purpose of his model was to provide a geometrical schematization of the complex planetary movements. Aristotle, however, could not be content with this. His aim was to find the connections between all the sets of spheres, thus making a single, uniform system of them all and presenting all the heavenly bodies, from the fixed stars to the moon and the centrally placed earth, as a single organic unit. In order to link each set of spheres with the one next to it, Aristotle was obliged to introduce more and more new spheres in between those already in existence, with the sole object of integrating all the links into a single system. The number of these connecting spheres amounted to 22, thus making a total of 55 spheres in Aristotle's model. The result of making the model paramount to the facts and turning it into an actual substance was to complicate the picture to such an extent that no trace of the original intention of simplification remained.

The model of concentric spheres continued to dominate the popular picture of the cosmos throughout the period from Aristotle to Copernicus. However, from as early as the second century B.C. onward, the Greek astronomers felt that it was not sufficiently flexible to meet the ever-increasing accuracy of observation. This progress showed that the velocities of the sun and the planets were not uniform and that this fact could not possibly be explained on the assumption that the earth was the centre of these movements. For all that, practically all the astronomers stuck to the geocentric theory. The contradiction was resolved simply by assuming that the stars really did move in circles (following the original Pythagorean-Platonic theory) and that these circles did inscribe the earth, as the geocentric theory demands, but that their centres did not coincide with the centre of the earth, i.e. the circles were eccentric. Before Kepler nobody thought of conceiving the courses of the planets as ellipses even though the geometry of conic sections, including the ellipse, had been developed in Greece as early as the second half of the fourth century B.C. and had reached its peak at the beginning of the second century in the work of Apollonius of Perga on conic sections. The circle was just as inseparable a part of the Greek cosmos as the straight line later became of the Newtonian cosmos.

In addition to the theory of eccentric circles, another attempt was made to explain the retrograde movements of the planets in the course of their revolutions around the earth, and in the scientific world this second theory displaced the concentric spheres' model. This was the famous theory of epicycles. Its originator is unknown, but Hipparchus already used it in combination with the eccentric circle theory; and Ptolemy included it, in its final formulation, in his great astronomical work. There is reason for thinking that Apollonius of Perga had already developed this theory. The epicyclic theory, like its rival, breaks down the characteristic movement of a planet into circles. According to it, the planet revolves in a secondary circle (epicycle) around a geometrical centre, which in its turn revolves in a primary circle around the earth or a centre near to the earth (if the movement is eccentric). Hence, for an observer on the earth, the combination of the planet's revolution in the secondary circle with the progress of this circle's centre along the perimeter of the primary circle appears as the characteristic "epicyclic" move-

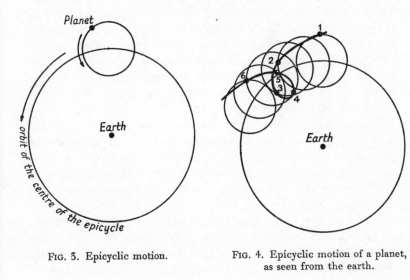

FIG. 3. Epicyclic motion.   FIG. 4. Epicyclic motion of a planet, as seen from the earth.

ment described above. To-day, after Copernicus, we know that the epicyclic movement in fact reflects the annual revolution of the earth around the sun. If the earth stood still, the other planets would seem to us to move along simple circles or ellipses.

It has already been remarked that, for geometrical reasons, the phenomenon is simpler in regard to the movements of the two "inner" planets, Mercury and Venus, which revolve between us and the sun. The centre of their epicycle is identical with the centre of the sun itself. The discovery of this fact was the first serious breach made in the geocentric theory. Heracleides of Pontus (388-315 B.C.), a pupil of Plato's, taught that Mercury and Venus revolve round the sun, while the sun, like all the other planets, revolves round the earth. We owe this information to late sources of the fourth and fifth centuries A.D.: "Lastly, Heracleides of Pontus, when describing the circle of Venus, as well as that of the sun, and giving the two circles one centre and one middle, showed how Venus is sometimes above, sometimes below the sun. For he says that the position of the sun, the moon, Venus, and all the planets, wherever they are, is defined by one line passing from the centre of the earth to that of the particular heavenly body. There will then be one straight line drawn from the centre of the earth showing the position of the sun, and there will equally be two other straight lines to the right and left of it respectively and distant 50° from it, and 100° from each other, the line nearest to the east showing the position of Venus or the Morning Star when it is furthest from the sun and near the eastern regions, a position in virtue of which it then receives the name of the Evening Star, because it appears in the east at evening after the setting of the sun" [117].

The second source observes more briefly: "Although Venus and Mercury are seen to rise and set daily, their orbits do not encircle the earth at all, but circle round the sun in a freer motion. In fact, they make the sun the centre of their circles, so that they are sometimes carried above it, at other times below it and nearer the earth, and Venus diverges from the sun by the breadth of one sign and a half" [118]. On this evidence Heracleides of Pontus may be considered the founder of the epicyclic theory which held the field, though without completely deposing the theory of concentric spheres, until the publication of Copernicus' book in 1543. Even then it did not pass into limbo: as is well known, Tycho Brahe at the end of the sixteenth century still tried to compromise between the geocentric and heliocentric theories by a hypothesis which was a kind of extension of Heracleides' theory. Brahe supposed that all the planets revolved round the sun, while the sun itself revolved round the earth.

Such was the astonishing power of the geometrical imagination of the ancient Greeks that the heliocentric theory appeared on the scientific horizon all those centuries ago. The hypotheses put forward included the rotation of the earth upon its axis as well as its revolution round the sun. But they failed to take firm root and were finally displaced by the geocentric theory which postulates the earth as the unmoving centre of the cosmos. The Pythagoreans were the first to deviate from this accepted doctrine. The theory of "the central fire", by displacing the earth from the centre, paved the way for other similar views. On this subject Aristotle writes as follows in his book *On the Heavens*: "It remains to speak of the earth, where it is, whether it should be classed among things at rest or things in motion, and of its shape. Concerning its position there is some divergence of opinion. Most of those who hold that the whole Universe is finite say that it lies at the centre, but this is contradicted by the Italian school called Pythagoreans. These affirm that the centre is occupied by fire, and that the earth is one of the stars, and creates night and day as it travels in a circle about the centre. In addition they invent another earth, lying opposite our own, which they call by the name of 'counter-earth', not seeking accounts and explanations in conformity with the appearances, but trying by violence to bring the appearances into line with accounts and opinions of their own. There are many others too who might agree that it is wrong to assign the central position to the earth, men who see proof not in the appearances but rather in abstract theory. These reason that the most honourable body ought to occupy the most honourable place, that fire is more honourable than earth, that a limit is a more honourable place than what lies between limits, and that the centre and outer boundaries are the limits. Arguing from these premises, they say it must be not the earth, but rather fire, that is situated at the centre of the sphere. . . . This then is the opinion of some about the position of the earth, and on the question of its rest or motion they are conformable views. Here again all do not think alike. Those who deny that it lies at the centre suppose that it moves in a circle about the centre, and not the earth alone, but also the counter-earth, as we already explained. Some even think it possible that there are a number of such bodies carried round the centre, invisible to us owing to the interposition of the earth. This serves them too as a reason why eclipses of the moon are more frequent than those of the sun, namely that it is blocked by each of these moving bodies, not only by the earth" [33].

The origin of this theory of the central fire is uncertain. Perhaps the discovery that the moon has no light of its own, but receives its light from the sun, led to the further conjecture that the sun too merely reflected the light of a central source, i.e. the central fire. This fire is hidden from our sight because the terrestrial globe in its revolution around it always presents its other uninhabited side to it. The counter-earth is likewise invisible to us, because it too is obscured by the earth during the revolution of these bodies round the sun, but the frequent eclipses of the moon are evidence of its existence. Such roughly, was the Pythagorean theory. From a historical point of view, its importance lies in the actual idea that the earth is not the centre of the universe but simply a star like any other. This conception opened the way for all the other similar theories. Herein lies its historical significance, which is in no way affected by the manner in which it came into being. In passing we may note that the Pythagorean scale of cosmic values, defining the centre as more "honourable" than any non-central position, reappears in a similar context in Copernicus' book when he argues that the earth moves and the heavens are at rest. "The condition of immobility is regarded more noble and divine than one of change and inconstancy. Hence movement should be attributed rather to the earth than to the universe. I would go further and say that it seems utterly incomprehensible to attribute motion to what contains and sustains and not to what is contained and sustained, i.e. the earth" (*On the Revolutions of the Heavenly Bodies*, I, 8). This similarity of argumentation is easily understood. Only Newtonian mechanics could provide physical reasons for the celestial movements; but this very mechanics only came into being as a result of the advances in astronomy and the theory of Copernicus. Therefore, in no small measure, Copernicus remained a contemporary of the Ancient Greek scientists.

In connection with the theory of the counter-earth, mention should be made of Aristotle's other explanation, which, like the first, is not altogether favourable to the Pythagoreans: "And all the properties of numbers and scales which they could show to agree with the attributes and parts and the whole arrangement of the heavens, they collected and fitted into their scheme; and if there was a gap anywhere, they readily made additions so as to make their whole theory coherent. E.g. as the number 10 is thought to be perfect and to comprise the whole nature of numbers, they say that the bodies which move through the heavens are ten, but as the visible bodies are only nine, to meet this they invented a tenth—the 'counter-earth'" [34]. This exposition is particularly interesting, since it contains no mention of the central fire; and the ninth of the visible bodies is apparently the sphere of the fixed stars, in accordance with Aristotle's theory. Now, there are certain indications that the later Pythagoreans restored the earth to its position as the centre of the cosmos and placed the fire in the centre of the earth. On the other hand, still later sources mention a new theory which was propounded by the Pythagorean School at the end of the fifth or the beginning of the fourth century B.C. and accepted by Heracleides of Pontus. "According to Theophrastus, Hicetas of Syracuse holds that the heavens, the sun, the moon, the stars and everything on high are fixed and stationary, and that, apart from the earth, nothing in the universe moves. But since the earth rotates upon its axis at a very great speed, everything appears as if she would be standing still while the heavens move" [35]. According to another tradition "Heracleides of Pontus and Ecphantus the Pythagorean move the earth, not however in the sense of translation, but in the sense of rotation, like a wheel fixed on an axis, from west to east, about its own centre" [119].

In the introduction to his book where he explains how he arrived at his theory, Copernicus specifically mentions the names of Hicetas, Ecphantus and Heracleides together with the ancient references to them. Here again we see the connection between the beginnings of modern astronomy and that of the fourth century B.C. There is an important difference, however: in the Ancient World the theory of the earth's rotation on its axis did not win a permanent place for itself, despite its profound influence on Plato. Admittedly, it is not clear whether Plato did in fact regard the earth as rotating: the crux in the *Timaeus* can be variously understood and has been differently interpreted. But there is no doubt that he considered the rotation of a sphere on its axis as the most perfect of all movements and ascribed it as a kind of universal movement to all the stars—including, possibly, the earth. For Plato, rotation on an axis was the supreme manifestation of reason, as he explains in the *Laws*: "Of these two motions, the motion which moves in one place must necessarily move always round some centre, being a copy of the turned wheels; and this has the nearest possible kinship and similarity to the revolution of reason. . . . If we describe them both as moving regularly and uniformly in the same spot, round the same things and in relation to the same things, according to one rule and system—reason, namely, and the motion that spins in one place (likened to the spinning of a turned globe)—we should never be in danger of being deemed unskilful in the construction of fair images by speech" [134]. Plato draws an analogy between reason and the perfectly precise movements of the heavens: "The whole course and motion of Heaven and of all it contains have a motion like to the motion and revolution and reckonings of reason" [133]. For the cosmos itself rotates on its axis, as Plato stresses in the *Timaeus*; that is to say, it lacks the three translational degrees of freedom (or, in Plato's language, the six movements up, down, forwards, backwards, right, left) and displays only the seventh movement—spin. Further on he reiterates that this motion is universal and is found in all the stars. It is amazing to observe how Plato, by allegorical reasoning and pure deduction, arrived at one of the empirical conclusions of modern astronomy and cosmology—namely, the universal existence of rotatory movement in the cosmos, as seen both in the rotation of the stars on their axes and in the rotations of larger systems and the cosmic galaxies.

Neither Plato's idea nor the hypothesis of Heracleides, Hicetas and Ecphantus was taken up by subsequent astronomers. The theory of spin was not an essential part of the Platonic definition of the stars as instruments of time. The circular motion of the stars round the earth, like any cyclical mechanism, is sufficient to act as a clock; and, indeed, it is the basis of our time measurement. The generally accepted opinion that rest is "nobler" than motion actually militated against the idea of universal rotation. No more striking example of this could be given than Plutarch's view on Plato's theory of time: "Therefore it is better to say that the earth is an instrument of time not in the literal sense, as having motion like the stars, but because by standing still it divides the risings and settings of the stars into periods by which the primary measures of time—day and night—are determined. . . . Just as pointers of sundials do not change their position with the shadows, but become instruments for measuring time by remaining in their place. They thus copy the earth which obscures the sun when the sun passes beneath it" [138]. The rejection of the theory that the earth rotates on its axis was eventually put on a "scientific" basis by Ptolemy, five hundred

years after Heracleides. Ptolemy's reasoning is entirely based on
ignorance of the law of inertia. and even Copernicus, living as he
did before the era of modern mechanics, could bring only qualitative arguments against it. Ptolemy's book is a summary of the
theories of previous generations, especially that of Hipparchus.
His attempt to disprove the views of Heracleides is thus of special
interest: "Certain thinkers, though they have nothing to oppose
to the above arguments, have concocted a scheme which they
consider more acceptable, and they think that no evidence can
be brought against them if they suggest for the sake of argument
that the heaven is motionless, but that the earth rotates about one
and the same axis from west to east, completing one revolution
approximately every day, or alternatively that both the heaven
and the earth have a rotation of a certain amount, whatever it
is, about the same axis, as we said, but such as to maintain their
relative situations. These persons forget however that, while, so
far as appearances in the stellar world are concerned, there might,
perhaps, be no objection to this theory in the simpler form, yet,
to judge by conditions affecting ourselves and those in the air
about us, such a hypothesis must be seen to be quite ridiculous. . . .
They must admit that the rotation of the earth would be more
violent than any whatever of the movements which take place
about it, if it made in such a short time such a colossal turn back
to the same position again, that everything not actually standing
on the earth must have seemed to make one and the same movement always in the contrary sense to the earth, and clouds and
any of the things that fly or can be thrown could never be seen
travelling towards the east, because the earth would always be
anticipating them all and forestalling their motion towards the
east, in so much that everything else would seem to recede
towards the west and the parts which the earth would be leaving
behind it. For, even if they should maintain that the air is carried
round with the earth in the same way and at the same speed,
nevertheless the solid bodies in it would always have appeared to
be left behind in the motion of the earth and air together, or,
even if the solid bodies themselves were, so to speak, attached
to the air and carried round with it, they could no longer have
appeared either to move forwards or to be left behind, but would
always have seemed to stand still, and never, even when flying
or being thrown, to make any excursion or change their position,

although we so clearly see all these things happening, just as if no slowness or swiftness whatever accrued to them in consequence of the earth not being stationary" [271].

Aristotle's part in the geocentric "ideology" will be discussed again below. The first Greek to adopt a thorough-going heliocentric standpoint was Aristarchus of Samos (c. 310-230 B.C.) who thus anticipated Copernicus by eighteen hundred years. His hypothesis that the earth revolves around the sun while rotating upon its own axis is worthy of the golden period of Greek science in which it was formulated. Aristarchus was a contemporary of Archimedes and Eratosthenes; he lived a few decades after Euclid and about a hundred years before Hipparchus and Poseidonius. We possess two important pieces of evidence about his theory. The first and weightier one, is Archimedes' reference to the same question in his *Sand-reckoner*, where he devises a method of expressing very large numbers and illustrates it by working out the number of grains of sand with a sphere the size of the cosmos. The relevant passage is as follows: "You are aware that 'universe' is the name given by most astronomers to the sphere the centre of which is the centre of the earth, and the radius of which is equal to the straight line between the centre of the sun and the centre of the earth; this you have seen in the treatises written by astronomers. But Aristarchus of Samos brought out a book consisting of certain hypotheses in which the premisses lead to the conclusion that the universe is many times greater than that now so called. His hypotheses are that the fixed stars and the sun remain motionless, that the earth revolves about the sun in the circumference of a circle, the sun lying in the middle of the orbit, and that the sphere of the fixed stars, situated about the same centre as the sun, is so great that the circle in which he supposes the earth to revolve bears such a proportion to the distance of the fixed stars as the centre of the sphere bears to its surface" [237].

Archimedes rightly observes that the last sentence does not make sense; for what relation can there be between the centre of a sphere, which is no more than a point, and its surface. Archimedes explains the words by supposing that Aristarchus assumed that the ratio of the terrestrial globe (pictured as a point) and the sphere containing the earth's orbit round the sun equalled the ratio of this latter sphere to the sphere of the fixed stars.

As regards the question which concerns us here, this passage from Archimedes mentions only the revolution of the earth round the sun. The second of our pieces of evidence also refers to the rotation of the earth upon its axis. It is found in Plutarch's book *On the Face in the Moon* which is a discussion between friends on the physical properties of the moon. "Do not bring against me a charge of impiety such as Cleanthes used to say that it behoved Greeks to bring against Aristarchus of Samos for moving the Hearth of the Universe, because he tried to save the phenomena by the assumption that the heaven is at rest, but that the earth revolves in an oblique orbit, while also rotating about its own axis" [235]. These words make it clear that Aristarchus' hypothesis failed to gain acceptance mainly for religious reasons. Three hundred years later Plutarch recalls the violent opposition of one of the founders of the Stoic School to the heliocentric theory and informs us that Aristarchus nearly met with the same fate as Anaxagoras, who was condemned for blasphemy because he taught that the stars were flaming stones. As a matter of fact, the position was still more serious in the time of Aristarchus than in Anaxagoras' day: the central position of the earth and its absolute rest, as formulated by Aristotle, had in the meantime become an axiom without in any way detracting from the divine quality of the stars in their unvarying courses. On the contrary, the two conceptions, being complementary and mutually support-ing, simply became two aspects of the same strongly religious cosmic view.

However, it would not be fair to dwell on the religious opposi-tion to the heliocentric theory while ignoring the scientific argu-ments adduced against it. We have already seen Ptolemy's reasons for rejecting the earth's rotation on its axis. They go back to the doctrine of Aristotle, who also did not accept the view that the earth revolves round the sun. All these arguments are based on the absence of a parallax in any of the fixed stars; they were finally refuted when it was proved empirically (in 1838 by Bessel) that such a parallax, though very small, does in fact exist. Even Copernicus could find no other answer to this argument than the assumption (which proved to be correct) that the parallaxes were too small to be measured. The answer given by Aristarchus is not known. Aristotle's objection is found in his book *On the Heavens*. There he justly remarks that, if the earth revolved round the centre, this movement would be reflected in the changes in the positions of the fixed stars. We may translate this into modern scientific terms as follows: just as the epicyclic movement of the planets reflects the annual revolution of the earth, so there must be an "epicyclic" movement of the fixed stars, i.e. a change in the angle (parallax) at which the earth appears, in the course of the year, to an observer at the distance of the fixed stars. Or, in Aristotle's language: "If this were so, there would have to be passings and turnings of the fixed stars. Yet these are not observed to take place: the same stars always rise and set at the same places on the earth" [160].

Much as we may admire Aristotle's acumen, we cannot but marvel still more at the scientific daring of Aristarchus' imagination. It is open to doubt if his heliocentric theory was in fact rejected only because of this argument about the parallax of the fixed stars; even if this parallax had been discovered in his day— or even in Copernicus' time—every effort would probably have been made to explain it in such a way as would have preserved the geocentric theory. Even Cleanthes' charge of blasphemy was simply a current version of a deeper antagonism. Copernicus and Galileo were in a very different position from Aristarchus. By their time the religious argument had been made immeasurably more weighty by the evidence of the Bible and the authority assigned to Aristotle by the Church. Aristarchus' conception was simply too bold for his contemporaries who were not yet ripe for the intellectual revolution involved in this relativization of the place of the observer in the cosmos.

It was a long road to the scientific boldness required for the acceptance of this relativization that finally led to Newton's mechanics and the breaking down of the antithesis between heaven and earth. On the other hand, Newtonian mechanics introduced new absolute values into the world of science and our conception of the cosmos—absolute space and absolute time. Therefore, it would be more correct to compare the conception of Aristarchus with that of Einstein in his restricted Theory of Relativity, which centres round the relativity of space and time. In Einstein's case, as in that of Aristarchus, the opposition to the abolishment of absolute notions sometimes appeared in scientific guise; and not a few scientists tried hard to explain the empirical evidence, such as the dependence of mass on velocity, by the Newtonian theory. Generally speaking, however, in our days of organized science, a new scientific theory is accepted by experts in the field if it is in accordance with all the known facts and opens the way to the discovery of new ones, even though it may involve changes in the structure of a whole philosophy. A contributory factor in the non-acceptance of Aristarchus' hypothesis was the isolation of the scientist in the Ancient World which resulted from the lack of organized tradition of scientific teaching. We know only of Seleucus, who taught his doctrine in Babylon in the second century B.C. Our source of information is Plutarch: "What does Timaeus mean by saying that there are souls scattered on the earth, the moon, and in the other instruments of time? . . . Did he mean to put the earth in motion as he did the sun, the moon, and the five planets, which he called the instruments of time, on account of their turnings, and was it necessary to conceive that the earth 'rolling about the axis stretched through the universe' was not represented as being held together and at rest, but as turning and revolving, as Aristarchus and Seleucus afterwards maintained that it did, the former stating this as only a hypothesis, the latter as a definite opinion? But Theophrastus adds to his account the detail that Plato in his later years regretted that he had given the earth the middle place in the universe which was not appropriate to it" [138].

The Greek picture of the cosmos was largely determined by the transition from seeing the sky in two dimensions, as a canopy stretched over the earth, to seeing it in three dimensions with the depth of space added. This process unfolded very rapidly from the moment when Pythagoras and his disciples discovered that the earth is a sphere. We are not told how they arrived at this discovery and it may well have started as pure conjecture. In Aristotle's book *On the Heavens* we find the very same proofs of sphericity as are used to this day in school geography: "If the earth were not spherical, eclipses of the moon would not exhibit segments of the shape which they do. . . . In eclipses the boundary is always convex. Thus if the eclipses are due to the interposition of the earth, the shape must be caused by its circumference, and the earth must be spherical. Observation of the stars also shows not only that the earth is spherical but that it is of no great size, since a small change of position on our part southward or northward visibly alters the circle of the horizon, so that the stars above our heads change their position considerably, and we do not see the same stars as we move to the North or South. Certain stars are seen in Egypt and the neighbourhood of Cyprus, which are invisible in more northerly lands, and stars which are continuously visible in the northern countries are observed to set in the others. . . . Mathematicians who try to calculate the circumference (of the earth) put it at 400,000 stades. From these arguments we must conclude not only that the earth's mass is spherical but also that it is not large in comparison with the size of the other stars" [161].

The numerical value given here by Aristotle is 50 per cent. too great, if we take the accepted length of the stade, i.e. 160 metres. We shall come back to these measurements at once. But first a quotation from the geographer Strabo (c. 63 B.C.-A.D. 18), who gives a further proof of the earth's shape: "The spherical shape of the earth is seen . . . in the phenomena of the sea and of the heaven. For this we have the evidence of the senses and of common sense. For it is obvious that the curvature of the sea hides from the sailors the distant lights which are at the level of their sight, whereas they can see the lights when they are above the level of their eyes even if they are further away. Similarly, from a high place the eye beholds what was hidden from it before. . . . Furthermore, as the sailors approach dry land more and more of the shore becomes visible and what at first seemed low rises progressively higher" [255].

The recognition of the earth's sphericity was accompanied from Aristotle onwards by the development of methods of measuring its circumference. A striking description of two such methods has been preserved in the *Theory of Revolutions of the Heavenly Bodies*, of Cleomedes, a Stoic who probably lived at the beginning of the present era. The basic principle in both methods is the same: measuring the distance between two points on the same meridian of the globe and determining the corresponding arc for this distance. One of the methods was used by Eratosthenes (c. 275-195 B.C.) at Alexandria, and is based on the altitude of the sun on the day of the summer solstice. The second was used by Poseidonius (c. 135-51 B.C.), the famous Stoic philosopher and

teacher of Cicero, who discovered the dependence of the tides on
the movement of the moon. Cleomedes describes Poseidonius'
method as follows: "Poseidonius says that Rhodes and Alexandria
lie under the same meridian. Now meridian circles are circles
which are drawn through the poles of the universe, and through

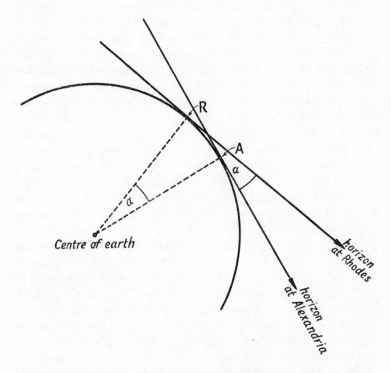

FIG. 5. Poseidonius' determination of the earth's circumference from
the distance between Rhodes and Alexandria and the angle between
their respective horizons.

the point which is above the head of any individual standing on the
earth. The poles are the same for all these circles, but the vertical
point is different for different persons. . . . Now Rhodes and
Alexandria lie under the same meridian circle, and the distance
between the cities is reputed to be 5,000 stades. Suppose this to
be the case . . . Poseidonius goes on to say that the very bright star
called Canobus lies to the south, practically on the Rudder of
Argo. The said star is not seen at all in Greece; hence Aratus does

not even mention it in his 'Phenomena'. But, as you go from north to south, it begins to be visible at Rhodes and, when seen on the horizon there, it sets again immediately as the universe revolves. But when we have sailed the 5,000 stades and are at Alexandria, this star, when it is exactly in the middle of the heaven, is found to be at a height above the horizon of one-fourth of a sign, that is, one forty-eighth part of the zodiac circle. It follows, therefore, that the segment of the same meridian circle which lies above the distance between Rhodes and Alexandria is one forty-eighth part of the said circle, because the horizon of the Rhodians is distant from that of the Alexandrians by one forty-eighth of the zodiac circle. . . . And thus the great circle of the earth is found to measure 240,000 stades, assuming that from Rhodes to Alexandria it is 5,000 stades; but, if not, it is in the same ratio to the distance. Such then is Poseidonius' way of dealing with the size of the earth" [241].

Assuming the ancient stade to be equal 160 metres, we find that the calculation of Poseidonius—as also the 250,000 stades arrived at by Eratosthenes—was within a few per cent. of the actual figure of 40,000 kilometres.

The first Greek to attempt any estimate of cosmic distances was Anaximander. It is related of him that he thought the circle of the moon nineteen times as great as the earth and that of the sun twenty-eight times as great [10]. The Pythagorean School, following its theory of the harmony of the spheres, supposed the ratio of the distances between the planets to be the same as the ratios of musical harmonies. The influence of this idea on Plato can be seen in the *Timaeus*, where he lays it down that the distances of the planets from the earth are in the ratio of $1 : 2 : 3 : 4 : 8 : 9 : 27$. Plato speaks of "double and triple distances", meaning apparently that these integers are a combination of two geometrical series, one with the factor 2 (1, 2, 4, 8) and the other with the factor 3 (3, 9, 27). In Plato's day the order of the planets in increasing distance from the earth was generally believed to be: the Moon, the Sun, Venus, Mercury, Mars, Jupiter, Saturn, During the third and second centuries the Greek astronomers adopted a new order based on the periods of the stars: Moon, Mercury, Venus, Sun, Mars, Jupiter, Saturn. This modification was probably influenced by Heracleides' hypothesis mentioned above, that Mercury and Venus revolve around the sun.

The first serious attempt to calculate astronomical distances or the ratios between them was made by Aristarchus. His methods and those of his followers mark the culminating point of the Greek achievements in mathematical astronomy. Aristarchus tried to calculate the relative distances of the sun and moon from the earth. He started from the assumption that, when the moon is half full, the earth, the moon and the sun form a right-angled

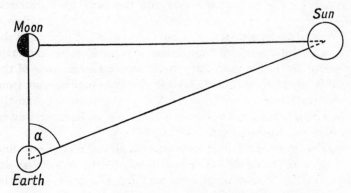

FIG. 6. Aristarchus' determination of the relative distances earth-sun and earth-moon.

triangle with its right angle in the moon. Therefore, to find the ratio of the side earth-moon to the hypotenuse earth-sun, we have to measure the angle between these two sides, i.e. the arc between the moon and the sun at the moment when the moon is exactly half illuminated. The difficulties in such a measurement are obvious, especially as it has to be taken by day at a time when both celestial bodies are above the horizon—a requirement which makes it still more difficult to determine the moment when the moon is exactly a semicircle. The method is not accurate, since the angle to be measured is itself almost 90° (89° 52′) and therefore a slight error in its determination produces a very large error in the result, i.e. in the ratio of the two distances.

The method and results are given in Aristarchus' essay "On the Sizes and Distances of the Sun and Moon", from which the following sentences are taken:

"1. That the moon receives its light from the sun.

2. That the earth is in the relation of a point and centre to the sphere in which the moon moves.

3. That, when the moon appears to us *halved*, the great circle which divides the dark and the bright portions of the moon is in the direction of our eye.

4. That, when the moon appears to us *halved*, its distance from the sun is then less than a quadrant by one-thirtieth part of a quadrant.

5. That the breadth of the earth's shadow is that of two moons.

6. That the moon subtends one-fifteenth part of a sign of the zodiac.

Given these hypotheses it is proved that:

1. The distance of the sun from the earth is greater than eighteen times, but less than twenty times, the distance of the moon from the earth: this follows from the hypothesis about the halved moon.

2. The diameter of the sun has the same ratio as aforesaid to the diameter of the moon.

3. The diameter of the sun has to the diameter of the earth a ratio greater than that which 19 has to 3, but less than that which 43 has to 6: this follows from the ratio thus discovered between the distances, the hypothesis about the shadow, and the hypothesis that the moon subtends one-fifteenth part of a sign of the zodiac" [236].

Aristarchus thus found the angle in question to be 87°, and this deviation of 2° 52' from the correct value produced a twenty-fold error in his determination of the distance of the sun in relation to that of the moon; this ratio is almost 400, not nearly 20 as Aristarchus held. Moreover, the apparent size of the moon is not 2° as he supposed, but approximately half a degree. These measurements were improved by Hipparchus and Poseidonius in the century after Aristarchus. They were completed by determining the distances of the moon and sun in units of the earth's diameter. The accuracy of all these figures varied from investigator to investigator, but in the end the distance of the moon was worked out with less than 10 per cent. of error, whereas the value for the sun's distance was brought no nearer than 60 per cent. of the actual figure and that of its diameter to 35 per cent.

Thus we find that the Greek astronomers arrived at the correct order of magnitude for some of the main distances within the solar system. However, it is not the numerical exactitude of their calculations that is important here but their use of geometrical and trigonometrical methods to measure celestial dimensions. The mathematization of the cosmos, the transformation of its expanses into a field of geodetic measurements to be handled in the same way as terrestrial dimensions, the recognition that space has depth and the determination of its distances in units of terrestrial geography—all this was the last link in the chain of Greek astronomical development which began with Pythagoras. This whole process was started by associating the precision of the celestial movements with the divinity of the stars. This association was so strong that it could not be broken by the constantly increasing accuracy of the methods employed or their affinity to secular, terrestrial measurements. We shall see later that the "atheism" of Epicurus, who tried to carry on the tradition of Anaxagoras and the Milesian rationalists by secularizing the heavens, failed to supersede the other tendency. The reason for this lay primarily in Epicurus' own method: his approach was less scientific, and the tradition of scientific accuracy remained inseparable from the "theistic" tradition both inside the great schools, such as that of the Stoics, and outside them. A scientist like Poseidonius regarded the sun as a divine force. The view of his pupil, Cicero, as expressed in his book *De Natura Deorum* is characteristic of the attitude of an intellectual in the last century B.C.: "Aristotle is also to be commended for his view that the motion of all living bodies is due to one of three causes, nature, force, or will; now the sun and moon and all the stars are in motion, and bodies moved by nature travel either downwards owing to their weight or upwards owing to their lightness; but neither is the case with the heavenly bodies, because their motion is revolution in a circle; nor yet can it be said that some stronger force compels the heavenly bodies to travel in a manner contrary to their nature, for what stronger force can there be? It remains therefore that the motion of the heavenly bodies is voluntary. Anyone who sees this truth would show not only ignorance but wickedness if he denied the existence of the gods" [173].

These words are most instructive, showing as they do how the sublimation of star-worship and the Pythagorean tradition were later reinforced by the scientific teachings of Aristotle. Of these we must now speak more fully.

79

IV

THE COSMOS OF ARISTOTLE

"For as the heavens are higher than the earth, so are my ways
higher than your ways, and my thoughts than your thoughts."

IS. 55.9

———————

ARISTOTLE'S physical doctrine was accepted as dogma for
sixty generations. No other personality in the history of
science, and very few in the whole course of human culture, had
so deep and long-lasting an influence on subsequent thought.
Already in the Ancient World Aristotle's views, or the views propounded in his name, bore the stamp of a supreme authority
which only a few bold spirits dared to reject. This authority was
in no way weakened by the rival claims made for Plato's philosophy at the end of the classical period and the beginning of the
Middle Ages. In the sphere of natural science there was no
essential difference of opinion between the two philosophies. The
chief interpreters of the three great religions finally blended the
main principles of Aristotle's philosophy with their religious
conception of the universe, thus turning the whole of that philosophy, including its physical and cosmic aspects, into unquestionable dogma.

This development was a serious obstacle to scientific progress
at the beginning of the modern period. But even in the Ancient
World it was Aristotle's dominating majesty that came to decide
whether scientific thought should advance or not. His influence
on physical science and the general picture of the cosmos was on
the whole more negative than positive. For this there were two
main reasons: Aristotle's approach to natural phenomena sprang

80

from a conception which held no key to the physical world, though it did have something to offer for the biological sciences in which Aristotle's achievements were greatest. Moreover, Aristotle tended to fit all his findings into fixed patterns and then to construct on them a general theory which he declared absolute. Whatever may be the methodological advantages of such an approach, they are outweighed by its disadvantages in a sphere which essentially depends on experiment and the nature of facts. Here dogmatic formulation becomes dangerous, especially if it precedes the collection of sufficient empirical evidence.

The guiding principle in Aristotle's view of nature was teleology: the axiom that everything that happens is done for a certain end and that the whole cosmos with all that it contains is the result of previous planning. Aristotle's philosophy is so dominated by the teleological principle that he used to be regarded as its inventor. In fact, it goes further back; only, he took it up, perfected it and gave it an importance that it had not possessed before. We saw earlier how Greek science began to search for a physical principle as the basis of phenomena. We found that it was Empedocles who first added to this principle the force (or, as he believed, the two opposed forces) that gives form to matter, binding it together and dividing it into parts through motion. This idea appeared again, in a somewhat modified form, in the doctrine of Anaxagoras. He, too, found it necessary to introduce into his picture of the cosmos a motive principle which he called "mind". The chief passage concerning the nature and functions of this "mind" may be variously interpreted: "Other things all contain a part of everything, but Mind is infinite and self-ruling, and is mixed with no thing, but is alone by itself. . . . Mind took command of the universal revolution, so as to make [things] revolve at the outset. And at first things began to revolve from some small point, but now the revolution extends over a great area and will spread even further. And the things which were mixed together, and separated off, and divided, were all understood by Mind. And whatever they were going to be, and whatever things were then in existence that are not now, and all things that now exist and whatever shall exist—all were arranged by Mind, as also the revolution now followed by the stars, the sun and moon, and the Air and

Aether which were separated off. It was this revolution which caused the separation off. And dense separates from rare, and hot from cold, and bright from dark, and dry from wet. There are many portions of many things and nothing is absolutely separated off or divided the one from the other except Mind" [57].

It would appear that Anaxagoras regarded "mind" as first and foremost the cause of motion, and also as a physical substance in essence. True, he describes it as distinct from all other substances, especially by virtue of its function in introducing order into the cosmos at the time of its formation and of maintaining that order now. But he nowhere mentions that the "mind" functions on a preconceived plan designed to produce a desirable, or good, or beautiful order. In the very term "mind", however, there was something that induced or accelerated such a process of thought. Evidence of this is to be found in the sentences of one of Anaxagoras' younger contemporaries, Diogenes of Apollonia: "Such a distribution would not have been possible without Intelligence, that all things should have their measure: winter and summer, and night and day and rains and winds and periods of fine weather; other things also, if one will study them closely, will be found to have the best possible arrangement" [69]. Here "intelligence" or "mind" is explicitly mentioned as a teleological agency that intentionally sets things in the proper order: the harmony of the universe did not come about of itself or by some physical principle, but through the force of an intelligently fashioning law which aims at perfection.

This development, which, as we see, began in the pre-Socratic period, reached its climax in the time of Plato. In the *Phaedo* Plato puts into the mouth of Socrates strong words of criticism against the "mind" of Anaxagoras, on the grounds that it has disappointed the hopes that its name led Socrates to place in it: "But one day I heard somebody reading from a book, by Anaxagoras as he said, the statement that Mind is the dispenser and cause of all things: and I was delighted with this cause, and it seemed to me in a certain way right that Mind should cause all things, and I reflected that, if this were so, then Mind in ordering all things must order and arrange them in the best possible way. If then one wished to discover the cause why anything comes into being or passes away or exists, he will have to discover how it is best for that thing to exist or to act or be acted upon in any way. . . . As I reasoned thus, I was happy to think that in Anaxagoras I had found a teacher after my own heart to explain the causes of things. . . . And I was ready to make similar enquiries about the sun and the moon and the other heavenly bodies with regard to their relative velocity, and the solstices and other conditions—just how it was better for each body to act and experience what it does. . . . My magnificent hopes were shattered, my friend, when, as my reading progressed, I found a man making no use whatever of Mind and ascribing to it no causal action in the ordering of things, but assigning such causes as air and aether and water and many other strange things" [121].

In the previous chapter this same charge against Anaxagoras appeared as Plato's attack on the view that the stars are flaming stones devoid of mind and soul. But the form given to it here in the *Phaedo* is of more general application. We may understand from the words of Socrates that he considers the teleological cause as the only true explanation of cosmic phenomena. While not denying the existence of other causes, he regards them as of no significance for the rational explanation of things, since they do not provide a reason for them nor connect them into a single orderly unit, all the parts of which are designed for a certain end. Plato's meaning is made quite clear in the famous passage in the *Phaedo*. "Is my sitting here in prison [says Socrates] explained merely by the fact that I have bones and joints and flexible muscles, all of which together enable me to sit here? Is not the true cause to be found rather in my decision to remain here after the verdict? Furthermore, are the words that I am uttering now explained solely by the laws of sound and the laws of hearing? There can be no doubt [he continues] that there is here a confusion of concepts between the true cause and the conditions which are necessary for its existence." The same is true of Plato's explanation of nature: men presume physical causes to explain the phenomena, without realizing that there is a force "which arranges them in the best possible way". To-day we need only look around us to see that the rapid advance of the natural sciences in recent centuries began only when scientists stopped searching for the "true causes" and confined their curiosity to the "necessary conditions" of those causes. The vistas of our modern cosmos were opened to us on the day when the scientists of the seventeenth century gave up asking "Why?" or "For what purpose?" and limited themselves to the question "How?"—to the investigation of "auxiliary causes" or secondary causes, as Plato calls them in the *Timaeus*.

In the *Timaeus* Plato rounds off his teleological theory with a cosmological allegory. There he describes in detail the mechanisms of the physical world which the demiurge uses for his ends, at the same time stressing that they are no more than causes which assist in the attainment of the best end. He then continues: "The great mass of mankind regard them, not as accessories, but as the sole causes of all things, producing effects by cooling or heating, compacting or rarefying, and all such processes. But such things are incapable of any plan or intelligence for any purpose. . . . And a lover of intelligence and knowledge must necessarily seek first for the causation that belongs to the intelligent nature, and only in the second place for that which belongs to things that are moved by others and of necessity set yet others in motion" [130].

The scale of values here imposed on the investigation of nature, culminating as it does in the teleological explanation of the universe, subsequently became the corner-stone of Aristotle's physics. Aristotle went still further and constructed his whole theory of dynamics on the supposition of an "intelligent nature" functioning by deliberate design. He defines his own stand in principle in the course of a refutation of those pre-Socratic philosophers who regarded natural phenomena as the product of "necessity", or in modern terminology of conformity to mechanical law. This necessity was supposed to be just as active now as it was when the cosmos came into being, when order was created out of primeval matter. Empedocles included the formation of the organic world too in this conjectured mechanism of necessity. In his view, nature created living creatures of all possible forms and in all possible combinations, including mixtures of man and beast. But not all these combinations survived: the forms which exist to-day were the only ones that could withstand the interplay of the forces of cosmic necessity.

For Aristotle all conformity to law is teleological, like that displayed in the artist's creation. The regular recurrence of natural phenomena is the manifestation of this type of law in which both plan and purpose are evident; whereas every deviation from regularity appears as a "freak", a mistake of the artist, as it were. In the second book of his *Physics*, Aristotle asks: "Why should not nature work, not for the sake of something, nor because it is better so, but just as the sky rains, not in order to make the corn grow, but of necessity? What is drawn up must cool, and what has been cooled must become water and descend, the result of this being that the corn grows. Similarly if a man's crop is spoiled on the threshing-floor, the rain did not fall for the sake of this—in order that the crop might be spoiled—but that result just followed. Why then should it not be the same with the parts in nature, e.g. that our teeth should come up of necessity— the front teeth sharp, fitted for tearing, the molars broad and useful for grinding down the food—since they did not arise for this end, but it was merely a coincident result; and so with all other parts in which we suppose that there is purpose? Wherever then all the parts came about just what they would have been if they had come to be for an end, such things survive, being organized spontaneously in a fitting way; whereas those which grew otherwise perished and continue to perish, as Empedocles says his 'man-faced ox-progeny' did. Such are the arguments (and others of the kind) which may cause difficulty on this point. Yet it is impossible that this should be the true view. For teeth and all other natural things either invariably or normally come about in a given way; but of not one of the results of chance or spontaneity is this true. We do not ascribe to chance or mere coincidence the frequency of rain in winter, but frequent rain in summer we do. Nor heat in the dog-days, but only if we have it in winter. If, then, it is agreed that things are either the result of coincidence or for an end, and these can not be the result of coincidence or spontaneity, it follows that they must be for an end; and that such things are all due to nature even the champions of the theory which is before us would agree. There- fore action for an end is present in things which come to be and are by nature. Further, where a series has a completion all the preceding steps are for the sake of that. Now surely as in intelli- gent action, so in nature; and as in nature, so it is in each action, if nothing interferes. Now intelligent action is for the sake of an end; therefore the nature of things also is so. Thus if a house, e.g. had been a thing made by nature, it would have been made in the same way as it is now by art; and if things made by nature were made also by art, they would come to be in the same way as by nature" [141].

This is a clear and unambiguous statement of the teleological view: nature's method is that of the artist, and conversely, true art is the imitation of nature. Accordingly, the scientist should approach his problem like the student of an artistic creation who, from the details of the house, learns the functions assigned by the builder to its various parts, or who understands from the shape of a statue what the artist wanted to express in it. This conception of natural phenomena as the striving for an end may be fruitful and valuable as a guiding principle in those sectors of biology where the subject of investigation is the functional role of organic forms and processes. Hence Aristotle's great achievements in zoology and the enduring value of no small part of his biological works. His treatises on the morphology of living creatures read as if they had been written by a contemporary of ours. Whereas his whole *Physics* is permeated by the spirit of a world entirely alien to us which began to pass away from the moment that physical science abandoned the teleological approach and replaced "For what purpose?" by "How?"

The teleological principle played a great part in Aristotle's dynamics. Since his theory of dynamics is an integral part of his picture of the cosmos, we must examine it in some detail. This will involve a description of his main ideas about the structure of the universe. From pre-Socratic physics Aristotle took over the theory of four elements and their division into heavy—earth and water, and light—air and fire. It occurred already to the earliest philosophers that there was a certain distribution of these elements in the cosmos: whereas earth and water are located around and in the earth, air belongs to a higher region and still higher is the fire concentrated in the stars and especially in the sun. Aristotle wove all these theories into the system of cosmic values drawn up by Plato on the basis of the antithesis between the eternal, well-ordered motion of the heavens and the irregular, ephemeral movements on earth. Aristotle completed this antithesis by emphasizing the essential difference in form between the movements of the stars and those in the lower region extending from the earth to the moon. The heavenly movements are circular, whereas those in our region are either more complex, like the trajectory of a missile, or take the form of a straight line, like the path of falling bodies (stones) or rising ones (vapour, flames). This scale of cosmic values postulates the priority of heavenly over terrestrial movements and the primacy of circular motion: "It can now be shown plainly that rotation is the primary locomotion. Every locomotion, as we said before, is either rotatory or rectilinear or a compound of the two: and the two former must be prior to the last, since they are elements of which the latter consists. Moreover, rotatory locomotion is prior to rectilinear locomotion because it is more simple and complete, which may be shown as follows. The straight line traversed in rectilinear motion cannot be infinite: for there is no such thing as an infinite straight line; and even if there were, it would not be traversed by anything in motion: for the impossible does not happen and it is impossible to traverse an infinite distance. On the other hand, rectilinear motion on a finite straight line is if it turns back a composite motion, in fact two motions, while if it does not turn back it is incomplete and perishable: and in the order of nature, of definition, and of time alike the complete is prior to the incomplete and the imperishable to the perishable. Again, a motion that admits of being eternal is prior to one that does not. Now rotatory motion can be eternal: but no other motion, whether locomotion or motion of any other kind, can be so, since in all of them rest must occur, and with the occurrence of rest the motion has perished. Moreover, the result at which we have arrived, that rotatory motion is single and continuous, and rectilinear motion is not, is a reasonable one. In rectilinear motion we have a definite starting-point, finishing-point, and middle-point. . . . On the other hand, in circular motion there are no such definite points: for why should any one point on the line be a limit rather than any other? Any one point as much as any other is alike starting-point, middle-point, and finishing-point" [152].

Aristotle's systematic mind would not allow him to associate this geometrically eternal movement with a substance which empirical evidence shows to move in a straight line in the sublunar region, viz. fire. Even if he agrees that the place of fire is in the higher regions, he holds that the substance of which the stars are made is of a superior kind which alone is capable of moving in a circle; that is to say, its natural form of motion is circular. The notion of a natural form of motion is connected with the concept of natural place, which we shall discuss in detail later. Now, it is clear that the natural place of earth and water is in the centre of the universe, on and in the earth. It is likewise clear that the natural place of air and fire is in the upper regions. The outcome of all this is the natural motion which results from the tendency of all the various kinds of matter which are not in their natural place to get to it, from their desire to occupy their proper place in the cosmos, thus correcting any deviation from its perfect order. In this way the natural movements become the dynamic manifestation of the working of teleology in the physical universe. To the original four elements a fifth has now to be added which, from the dynamic standpoint, is merely a third category, since the other four primary elements pair off into two sets of natural movement: "The same can also be proved on the further assumption that all motion is either natural or unnatural, and that motion which is unnatural to one body is natural to another, as the motions up and down are natural or unnatural to fire and earth respectively; from these it follows that circular motion too, since it is unnatural to these elements, is natural to some other. Moreover, if circular motion is natural to anything, it will clearly be one of the simple and primary bodies of such a nature as to move naturally in a circle, as fire moves upward and earth downward. . . . If fire be the body carried round, as some say, this motion will be no less unnatural to it than motion downwards. . . . Thus the reasoning from all our premises goes to make us believe that there is some other body separate from those around us here, and of a higher nature in proportion as it is removed from the sublunary world. . . . It seems too that the name of this first body has been passed down to present time by the ancients, who thought of it in the same way as we do . . . thus they, believing that the primary body was something different from earth and fire and air and water, gave the name 'aither' to the uppermost region. . . . It is also clear from what has been said why the number of the simple bodies, as we call them, cannot be more than we have mentioned. A simple body must have a simple motion, and we hold that these are the only simple motions, circular and rectilinear, the latter of two sorts, away from the centre and towards the centre" [153].

The last sentences of this passage are typical of Aristotle's physics; once a general formula has been decided on, there can be no argument: everything is definitely fixed, as for example that there is no place for another "simple body" in our cosmos. It may be argued that modern physics, too, proceeds in the same way and that the transition to relativistic mechanics involved a complete mental revolution, just because the formulae of Newtonian physics had become theoretical dogma. But there is an essential difference between Newton and Aristotle. Experiment plays a far greater part in modern physics, so much so that it is now the final arbiter of every theory. Hence science of our day has assumed a much more flexible character, its continuous progress being made possible by constant re-evaluation of its fundamentals. Aristotle's approach, on the other hand, failed to observe the right balance between induction and deduction and came to be dominated largely by the latter. His dogmatism and inclination for classifying at any cost petrified science by depriving it of flexibility and thus, in the absence of the corrective of experimental checks, blocked the road to further development. The psychological result of this way of thinking was that wherever experiment and theory were in conflict, it was the experiment that was found to be faulty.

Since the stars are made of aether which has a natural circular movement, how is the heat of the sun to be explained? Aristotle's answer to this question is quite clear: "The heat and light which they [the stars] emit are engendered as the air is chafed by their movement. It is the nature of movement to ignite even wood and stone and iron, *a fortiori* then that which is nearer to fire, as air is. Compare the case of flying missiles. These are themselves set on fire so that leaden balls are melted, and if the missiles themselves catch fire, the air which surrounds them must be affected likewise. . . . But the upper bodies are carried each one in its sphere; hence they do not catch fire themselves, but the air which lies beneath the sphere of the revolving element is necessarily heated by its revolution, and especially in that part where the sun is fixed" [157].

We shall now turn to the other two natural movements—upward and downward. Aristotle made these two directions into absolute opposites, in this disagreeing with Plato and others who understood that, in a universe which has a centre and a spherical perimeter, such concepts must be relative. In the same way and corresponding to these geometrical concepts, Aristotle also turned "heavy" and "light" into absolutes. This was the origin of that Aristotelian antithesis which so greatly hampered the formation of the fundamental notion of "specific weight" : "Our next topic must be weight and lightness. What are they, what is their nature, and why have they their particular powers? This is an enquiry relevant to a study of motion, for in calling things heavy and light we mean that they have the capacity for a certain natural motion. . . . Weight and lightness are predicated both absolutely and relatively. That is, of two heavy things we may say that one is lighter and the other heavier, e.g. bronze is heavier than wood. Our predecessors have treated of the relative sense, but not of the absolute: they say nothing of the meaning of weight and lightness, but discuss which is the heavier and which the lighter among things possessing weight. Let me make my meaning clearer. There are certain things whose nature it is always to move away from the centre, and others always towards the centre. The first I speak of as moving upwards, the second downwards" [163].

Here Aristotle refers to Plato's attempt to distinguish the various degrees of heaviness in "heavy" bodies, i.e. bodies with a natural downward tendency. Theophrastus tells us that Plato identified heaviness with weight: "He defined heavy and light not absolutely, but with reference to bodies such as earth. It would appear that of such bodies the heavy is with difficulty moved to a place not its own, whereas the light is moved easily" [137]. Since the "place not its own" of a heavy body is above, it follows that Plato discovered a simple and correct criterion for differentiating weights—namely, lifting the body. Aristotle, so far from attaching any special importance to this definition, congratulates himself on having "realized" that heavy and light are absolutes—one of those completely unfounded theories that did so much harm to the progress of physics. For Aristotle, the establishment of this absolute antithesis was a great achievement, because it completed his theory of matter which was built upon a combination of opposites. Having rejected the monistic theories of the Milesian and atomic schools in favour of the four elements of Empedocles, with the addition of the "eternal" aether, he had

90

to find a common basic principle for them all as a substitute for the monistic idea which derived everything from a single source. This he found in the principle of opposites which he made up out of the two pre-Socratic antitheses, hot-cold and dry-moist. From all their possible combinations he builds up the four elements in such a way as to give the 'equations': cold+dry= earth, cold+moist=water, hot+moist=air, hot+dry=fire. The first two equations produce the quality "heavy" which is the opposite of the quality "light" produced by the other two. In this way Aristotle replaces the four elements by the four qualities, hot, moist, dry and cold. Then, by combining and interchanging them with their opposites, he finally reaches his goal, which is to prove that the elements have a common source and can change into each other: "We maintain that fire, air, water and earth are transformable one into another, and that each is potentially latent in the others, as is true of all other things that have a single common substratum underlying them into which they can in the last resort be resolved" [166].

In the second part of his book *On Generation and Corruption* Aristotle propounds the main principles of his theory of matter. There he explains the transition from element to element by the interchange of qualitative factors in a given combination. Obviously fire is produced from air when moisture is changed into dryness, and earth from fire if heat is replaced by cold, and so on. Aristotle similarly examines other pairs of opposites such as hard-soft, rough-smooth, dense-rare and their role in the differentiation of the primary bodies. The history of physics has proved that all this theory of absolutely opposed qualities, even when presented with dialectical brilliance in the form of thesis and antithesis, leads nowhere. It has been shown that a far more productive method of describing natural phenomena is to define a given physical quality in terms of a continuous scale of numerical values, passing by gradations from one extreme to the other and doing away with the whole conception of absolute antithesis. The thermometer, for example, measures the degree of heat in a body under all conditions, thus reducing the subjective antithesis hot-cold to an objective scale of values. The same applies to the hygrometer, which measures all the degrees of moisture from maximum dryness to extreme humidity. Hard and soft cease to be opposites, once hardness has been defined and a method of measurement worked out from this definition which allocates to every body a certain place on a continuous scale of "degrees of hardness", etc. In modern times, the theory of opposites led Goethe astray, when he tried to maintain against Newton his own theory of colour which was based on the antithesis light-shade. It was no mere chance that Goethe disapproved of the mathematization of nature which involved divorcing the description of nature from man's subjective comprehension; nor is it surprising that Hegel, who fought so bitterly against Newton's views, should have approved so heartily of Goethe and his definition of colour as a synthesis of light and darkness.

We shall now return to our main problem—Aristotle's dynamics. The study of this question is in itself Aristotle's chief claim to fame, for he was the only ancient scientist who worked on the development of a quantitative or semi-quantitative theory in this fundamental field. This was no simple matter: the mechanics of Galileo and Newton have shown us that the phenomena of dynamics, which seems elementary to a layman, are in fact very complicated. We know now that the study of the laws of motion is complicated by two factors—friction and resistance of the environment—which cannot easily be neutralized. On account of these factors, the velocity of a body constantly decreases, after the increase resulting from the initial impulse. This fact gave rise to the "pre-Galilean" misconception that, in order to maintain a constant velocity, the body has to be subjected to a constant force; actually, such force merely serves to overcome the opposing forces of friction and of the resistance of the medium. Were it not for these opposing forces, the action of a constant force upon the body would steadily increase its velocity; while the action of a force which is only applied at the beginning of the motion would result in the body's permanently maintaining the velocity it had attained at the moment when the force ceased to act upon it. The problem is further complicated by the fact that we are within the earth's gravitational field. Thus, if we throw an object, its trajectory depends also on the force of gravity and assumes a shape which is the product of several factors, e.g. the inertia of the object, the effect of gravity and the resistance of the air.

All these findings of modern physics are the result of experimental and theoretical analysis of the various factors operative at the time of movement. Since Aristotle did not break down motion into its component factors, it is not surprising that his laws of motion do not correspond to the facts: "If, then, A the movent have moved B a distance C in a time D, then in the same time the same force A will move $\frac{1}{2}B$ twice the distance C, and in $\frac{1}{2}D$ it will move $\frac{1}{2}B$ the whole distance C: for thus the rules of proportion will be observed . . ." [151]. In other words, Aristotle maintains that the distance traversed by a body is in direct proportion to the force (which acts upon it constantly) and the time taken, and in inverse proportion to the body's mass. But experience showed him that this law was only a very rough approximation and that, when the disproportion between force and mass is very great, it fails completely. He therefore found it necessary to qualify his law and to give empirical examples in support of the qualification: "But if E move F a distance C in a time D, it does not necessarily follow that E can move twice F half the distance C in the same time . . . in fact it might well be that it will cause no motion at all; for it does not follow that, if a given motive power causes a certain amount of motion, half that power will cause motion either of any particular amount or in any length of time: otherwise one man might move a ship, since both the motive power of the shiphaulers and the distance that they all cause the ship to traverse are divisible into as many parts as there are men" [151]. In this passage we hear an entirely different Aristotle, an Aristotle who is far closer to us than the one who dogmatizes about the structure of the cosmos, never for one moment doubting the analytical power of pure intellect. Although his law of motion, quoted above, is not correct, its formulation is, in two respects, akin to the spirit of our own time: it is mathematical in form, and its validity is qualified by reference to experience.

Aristotle's law of motion applied to every movement: not only to "natural" movement, but also to "compulsory" movement in another direction than up or down. All the same, Aristotle formulated it again when he came to the subject of natural movement and thus included in it the motion which we call free fall. In this case, as with the upward movement of "light" bodies, the motive force is replaced by weight and its Aristotelian

93

opposite—"lightness". The various formulations show that in these cases, too, Aristotle was careful not to infringe the principles of his general law: "The larger a body the more swiftly it performs its proper motion" [158]. "The larger quantity of fire or earth always moves more quickly than a smaller to its natural place" [156]. "The large quantity (of fire) moves upwards more quickly than the small. Similarly a larger quantity of gold or lead moves downwards faster than a smaller, and so with all heavy bodies" [164].

Finally, to complete the picture, let us quote the quantitative formulation: "If a certain weight move a certain distance in a certain time, a greater weight will move the same distance in a shorter time, and the proportion which the weights bear to one another, the times too will bear to one another, e.g. if the half weight covers a distance in x, the whole weight will cover it in $\frac{1}{2}x$" [155]. Aristotle's law of falling bodies, then, says that the velocity of a falling body is proportional to its weight. This law became widely known through the deadly criticism levelled against it by Galileo in his book *Discourses and Mathematical Proofs* (published in 1638). Since then it has been an accepted fundamental of mechanics which has been proved experimentally that the velocity of fall is the same for all bodies and does not depend on their weight. This law is literally true only in a vacuum; but the modifications caused, for example, by the resistance of air are not very great, as long as the heights involved are not too large nor the weights too small. Certainly, there is nothing like the proportional relation postulated by Aristotle. Galileo rightly questions whether Aristotle ever carried out an experiment to see "if a stone falling from a height of 10 cubits reaches the earth at the same time as another stone ten times as heavy falling from a height of 100 cubits".

Galileo's criticism of this law marked a turning-point in the transition from ancient to modern physics. However, it is not enough for us to state the fact that Aristotle was mistaken: we must examine the problem in more detail. The whole subject of the laws of falling bodies is of significance outside the narrow limits of the law of a specific movement, through its connection with Aristotle's absolute negation of a vacuum and with his view about the influence of the environment on motion. It thus closely concerns the main principles of his conception of the physical world.

The founders of the atomic school (who form the subject of the following chapter) regarded the existence of a vacuum as a necessary condition of movement. Their conception was radical in the extreme: the ultimate elements of matter must be separated from each other by the absence of matter, i.e. by a vacuum which does not contain matter even in its most rarefied form, such as air or aether. Every atom moves in a vacuum until it collides with another atom. Thus, in the world of atoms, all activity is made up of movements in an absolute vacuum and impacts of matter on matter. This, one of the profoundest conceptions of Ancient Greece, was developed in the main by theoretical argument and remained unconfirmed by experiment until modern times. Aristotle's rejection of it was also an objection in theory and on principle; at least the theoretical objections have more weight in terms of his cosmic conception than his empirical arguments. The gist of these latter is mainly that changes in the volume of a body resulting from contraction or expansion are not proof of the existence of a void, since they do not conflict with the continuity of matter: contraction, for instance, is explained by the release of air from within the compressed body, and expansion by qualitative changes affecting a continuous substance, as in the case of evaporation. Since systematic experimentation was most imperfect in the Ancient World, it was the theoretical discussion, and not empirical arguments, that decided the principles to be adopted. Amongst the many reasons advanced by Aristotle against the existence of a vacuum, in the fourth part of his *Physics*, the most characteristic of his way of thinking is the following sentence: "The void, in so far as it is void, admits no difference" [142]. By this he means that the vacuum has no distinguishing geometrical features, nor any of the qualities essential for the fixing of direction and motion; in a vacuum there is no means of spatial orientation.

There is an analogy here with the negation of absolute motion in Newtonian space. If there were only one single body in the whole emptiness of infinite space, there would be no sense in assigning to it a place or a state of motion. For that, at least two bodies are required, thus defining relative motion in a given frame of reference. Aristotle sees no way of constructing such a framework if bodies are separated from each other by a vacuum. Since no geometrical link connects a body surrounded by the void to another body, there is no point in depicting its state either in terms of place or movement. Since Newton, physicists have grown accustomed to regard space as a geometrical network spread over the vacuum and joining together the physical points dotted about it. Aristotle rejected physical description by an abstract geometry extending beyond the bounds of matter or into the emptiness between its parts. Instead, he identified space with the volume filled by matter, an identification which necessitates the continuity of matter. It is characteristic of such a conception that Aristotle does not use the word "space", but "place", to express the location of a given body. "Place" is a far more concrete term than "space"; it gives a geometrical definition of a particular body in terms of the boundaries between it and its material environment, i.e. between it and the body or bodies which are in direct contact with all its periphery. Aristotle's combination of geometry and matter to form his concept of place is not unlike the conception of space in the General Theory of Relativity. This theory also rejects the Newtonian portrayal of space as a sort of infinite "box" in which physical bodies move. Instead, it pictures space as a kind of communion of the body and its surroundings: it is the body that determines the geometry of its environment, and this geometry cannot be artificially separated from the body itself. Hence a physical point is simply a singularity in the "metric field" which surrounds it. Again, this field is not at all an empty space, but a kind of emanation of the matter in it, just as matter is a kind of "materialization" of the field. This reasoning, like Aristotle's, also leads to the negation of a vacuum. The cosmos as envisaged by us to-day is very different from the "empty box" of Newton or of the Greek atomists. Interstellar space is full of electromagnetic radiation of every wave-length; its expanses contain gravitational fields and are traversed by gravitational waves. Similarly, there are fields of force round the atoms of which physical bodies are composed; and in the spaces within the atom or nucleus there is likewise interplay of forces at work between the primary particles.

This resemblance between Aristotle's approach to the space-continuum problem and that embodied in the General Theory of Relativity is very interesting, but we should beware of drawing

sweeping conclusions from it. Their similarity is offset by a fundamental difference. Whoever adopts the theory of relativity may in
every actual case fall back on Newton's theory as a first approximation to reality, an approximation which is frequently quite
sufficient for the description of the facts. In spite of the theoretical
and philosophical difference between classical mechanics and the
new theory, and in spite of the formal difference in their mathematical method, the former is still included in the latter as a first
approximation. History of science in the past three hundred years
is characterized by a chronological and almost organic sequence
in its development which was not found to anything like the
same extent in Greek science. It is above all the history of physics
from Galileo to our time that makes us realize that science
advances towards reality so to say by concentric approximations—
each theory containing its forerunner as a "special case". On the
other hand there is nothing to bridge the gulf, e.g. between
Aristotle's conception and that of Democritus before him. On the
whole, one would rather say that many of Aristotle's ideas, as
much as one admires their intellectual acumen, were in the
nature of a regression from those of the early atomists, since they
checked the development of certain sound principles in the
doctrine of the latter.

We must consider Aristotle's reasoning about falling bodies in
the light of his negation of a vacuum on the ground that it allows
of no physical action. On his teleological conception, the tendency
of a body to return to its natural place must necessarily increase
as its mass increases; in other words, the velocity of a falling
"heavy" body must increase with its weight, and the velocity of
the upward movement of a "light" body must increase with its
"lightness". This conclusion can be fitted into his general law of
motion, if the "tendency" of the body is regarded as the force
that sets it in motion: "We see that bodies which have a greater
impulse either of weight or of lightness, if they are alike in other
respects, move faster over an equal space, and in the ratio which
their magnitudes bear to each other. Therefore they will also
move through the void with this ratio of speed. But that is impossible; for why should one move faster? . . . Therefore all will
possess equal velocity. But this is impossible" [145]. Aristotle
here applies the principle of the lack of a sufficient reason to the
vacuum which allows of no differentiation. In this way he arrives

97

at the rejection of a conclusion which to-day we know to be correct—namely, that in a vacuum all bodies fall with the same velocity.

The negation of a vacuum is so important to Aristotle that he discusses the question again and again from various angles: "All movement is either compulsory or according to nature, and if there is compulsory movement there must also be natural (for compulsory movement is contrary to nature, and movement contrary to nature is posterior to that according to nature, so that if each of the natural bodies has not a natural movement, none of the other movements can exist); but how can there be natural movement if there is no difference throughout the void or the infinite? For in so far as it is infinite, there will be no up or down or middle, and in so far as it is a void, up differs no whit from down; for as there is no difference in what is nothing, there is none in the void (for the void seems to be a non-existent and a privation of being), but natural locomotion seems to be differentiated, so that the things that exist by nature must be differentiated. Either, then, nothing has a natural locomotion, or else there is no void" [143]. The commonest example of "compulsory" motion is the path of a projectile. Since in this case apparently the motive force acts only at the moment of throwing, Aristotle holds that the body is subsequently kept in motion by the environment, i.e. the air: "Further, in point of fact, things that are thrown move though that which gave them their impulse is not touching them, either by reason of mutual replacement, as some maintain, or because the air that has been pushed pushes them with a movement quicker than the natural locomotion of the projectile wherewith it moves to its proper place. But in a void none of these things can take place, nor can anything be moved save as that which is carried is moved" [143]. In summing up his arguments Aristotle mentions, only to reject, a conclusion which is in fact an explicit formulation of Galileo's law of inertia: "Further, no one could say why a thing once set in motion should stop anywhere; for why should it stop here rather than here? So that a thing will either be at rest or must be moved *ad infinitum*, unless something more powerful get in its way" [143].

However, Aristotle is not content with examining the problem only from one side. In the case of projection he laid stress on the subsidiary part played by the environment, without which he is unable to explain how there can be movement apparently without impulse. In the case of falling he similarly finds the retarding factor in the environment. Once again he puts forward a theory which is unsupported by experience: the velocity of fall is in inverse proportion to the density of the retarding medium. This also leads him to the conclusion that a vacuum is impossible: "Further, the truth of what we assert is plain from the following considerations. We see the same weight or body moving faster than another for two reasons, either because there is a difference in what it moves through, as between water, air, and earth, or because, other things being equal, the moving body differs from the other owing to excess of weight or of lightness. Now the medium causes a difference because it impedes the moving thing, most of all if it is moving in the opposite direction, but in a secondary degree even if it is at rest; and especially a medium that is not easily divided, i.e. a medium that is somewhat dense. *A*, then, will move through *B* in time *C*, and through *D*, which is thinner, in time *E* (if the length of *B* is equal to *D*), in proportion to the density of the hindering body. For let *B* be water and *D* air; then by so much as air is thinner and more incorporeal than water, *A* will move through *D* faster than through *B*. Let the speed have the same ratio to the speed, then, that air has to water. Then if air is twice as thin, the body will traverse *B* in twice the time that it does *D*, and the time *C* will be twice the time *E*. And always, by so much as the medium is more incorporeal and less resistant and more easily divided, the faster will be the movement. Now there is no ratio in which the void is exceeded by body, as there is no ratio of nullity to a number. . . . Similarly the void can bear no ratio to the full, and therefore neither can movement through the one to movement through the other, but if a thing moves through the thickest medium such and such a distance in such and such a time, it moves through the void with a speed beyond any ratio" [144].

The negation of a vacuum results therefore from the absurd conclusion that the velocity of fall in a vacuum must be infinite; the way Aristotle arrives at it is by "transition to the limit" from a medium of finite density to one whose density is nil. All Aristotle's conceptions of the nature of "place" rule out the physical possibility of such a transition. Just as the division of a body or of a section is an infinite process without limit, so the process of rarefication never ends and an infinite gulf separates the most rarefied medium imaginable from the vacuum. The former still allows of direction and movement, whereas in a vacuum they have no meaning. If the cosmos is a perfect continuum, it must be a single entity, a uniform body. Aristotle holds that this body is finite. While he assumes the possibility of infinity in division—indeed his conception of a continuum obliges him to adopt such a view—he denies it in extension. The cosmos is thus finite. Further, since the demand for perfection requires and astronomical phenomena prove its sphericity, the cosmos is a finite sphere with the earth as its centre and the sphere of the fixed stars as its limits. The finiteness of the cosmos also follows from its circular motion; otherwise we should have to admit an infinite velocity: "Again, if the heaven be infinite, and revolve in a circle, it will have traversed an infinite distance in a finite time. Imagine one stationary heaven which is infinite, and the other moving within it and of equal extent. Should the moving heaven, being infinite, have completed the circle, it will have traversed its equivalent, the infinite, in a finite time. But this we know to be impossible" [154].

The idea that the cosmos is finite is a corollary of Aristotle's dynamics and his conception of natural place. Just as light bodies tend to the limits of the cosmos, so heavy bodies tend to its centre, which is their natural place, and only a finite body can have a centre. Further, since the earth itself is composed of the heavy elements, the heavy bodies tend to the earth's centre, which means that this point is identical with the centre of the cosmos. Now, it is clear from all this that the earth must be in a state of absolute rest. The whole order of the cosmos, and particularly the circular movements of the celestial spheres and the stars attached to them, being eternal makes the cosmos itself eternal. Unlike the pre-Socratic philosophers, Aristotle rejects the idea of creation, since what is created must eventually decay. Therefore the eternity of the cosmos extends in the two temporal directions, past and future. It goes without saying that Aristotle criticizes Plato for basing his cosmology on the creation of the world and at the same time holding that the world will last for all futurity. The two great philosophical schools which came into being after Aristotle did not take up these views. Instead, the Stoics and Epicureans returned to the idea of cosmic development. Aristotle's theory about the eternity of the cosmos thus remained a unique, individual opinion throughout ancient times.

The two fundamentals of Aristotle's natural philosophy were developed and modified in the centuries following his death in the physics of the Stoics. The teleological idea appeared in their teaching in its most downright religious form as Providence. This concept preserves the supposition that everything is for the best and thus complements and softens the unbending rigour of the Stoic notion of fate, i.e. the absolute rule of causality. The continuum view of the cosmos took on an extreme dynamic form in the Stoic School. It emerged as the doctrine that all the parts of the cosmos are interdependent and that it is this inter-dependence that makes the cosmos a field of physical activity which pervades it and unites it into a dynamic whole. Aristotle's assumptions about the finitude of the cosmos and his absolute negation of a vacuum were not accepted in their entirety until the Middle Ages. The Stoics postulated the existence of an infinite vacuum surrounding the finite continuum of the cosmos; while Epicurus and his school took up and developed the theory of Leucippus and Democritus about the atomic structure of matter and the infinite extensions of the vacuum in the space between and outside the atoms. This deviation from the Aristotelian conception, though shortlived, also brought about a change of opinion with regard to falling bodies. We shall see below that the Epicureans held that the atoms fall through the void at the same speed, regardless of their weight; and one of the later commen-tators of Aristotle, John Philoponus, who lived at the beginning of the sixth century A.D., disagrees with his argument that, in the absence of a retarding medium, the velocity of fall would reach infinity. The attitude of Philoponus to Aristotle's dynamics is critical through and through; it is the last word of such criticism for the thousand years which preceded Galileo. Some of Philoponus' reasoning, especially his reflections on the relation of Aristotle's findings to experience, could serve as an introduction to Galileo's polemic.

There are many indications that a critical attitude to several of Aristotle's fundamental ideas made itself felt in the Hellenistic period. To take one example: from one of Plutarch's works, which deals with the physical properties of the moon, it is clear that such fundamental Aristotelian concepts as natural place and natural movement had begun to be challenged round about the beginning of the Christian era. The first signs appear of a conception of motion which is close to our own, especially in its resolution into components and the appeal to empirical analogies. This tendency was fostered by various factors, apart from the Epicurean and Stoic Schools—in so far as the latter deviated from Aristotle. First to be borne in mind is the progress made by astronomy in the third and second centuries B.C. which resulted in the discovery of facts entirely unknown to Aristotle, such as the precession of the equinoxes. This discovery of Hipparchus added to the cyclical movements in the heavens a new one which was not mentioned in Aristotle's cosmos. This in itself was no doubt sufficient to shake the blind faith in the validity of Aristotle's method and authority. A similar, by no means negligible, effect was produced by the development of ballistics and applied mechanics following the expansion of the Roman Empire and the growth of its military power. The study of projectile trajectories led to greater comprehension of the laws of dynamics. It became clear that the momentum of missiles depended only on the force applied to them at the moment of discharge and that Aristotle's theory about the impulse imparted to the stone during its flight by the air was erroneous. This was specifically demonstrated by John Philoponus, whose findings no doubt summed up the empirical research of the long period commencing with Archimedes and the first developments in the technique of war machines. Philo of Byzantium, Hero of Alexandria and the Roman Vitruvius had the true engineer's way of looking at things. They therefore improved their machines by empirical methods, such as the study of friction and the attempt to reduce it, or the analysis of the connection between applied forces and the movements resulting from them. In so doing, they established the basic principles underlying simple machines, and also some fundamental laws of motion, such as the law of the parallelogram of velocities. In consequence of all these achievements, Aristotle's dynamics was placed in its true historical perspective as a theory which was important as the beginning of a long development, though several of its concepts were subsequently proved wrong by experiment.

The teleology of Plato and Aristotle did not go unchallenged even in the Ancient World. The atomic theory of Democritus, which became part of Epicurean doctrine, was based on an entirely non-teleological approach. However, the Epicurean School had a much less lasting influence than its rival, the Stoic School, which upheld the teleological conception. Another keen critic of teleology was Aristotle's great pupil, the founder of systematic botany, Theophrastus. In his *Metaphysics*, Theophrastus disputes the view that the celestial movements are a special phenomenon essentially different from terrestrial ones, including those of living creatures: "Motion is a characteristic of all nature and especially of the heavens. Therefore, if activity is of the essence of everything and each thing, when active, moves as in the case of living creatures and plants (otherwise their names would be homonymous), then it is clear that it is also the essence of the heavens to revolve. For if they were separate and at rest, their name would be homonymous. For the revolving of the cosmos is a kind of life" [175]. Theophrastus is here challenging Aristotle's theory that the celestial spheres are "divine" entities capable of maintaining the eternal motion of the stars without any external impulse. Despite his anti-mechanistic outlook, Theophrastus here joins forces with the atomists who regarded this motion as one of the fundamental data of the universe, like matter, the reason for which it is pointless to seek. This criticism leads Theophrastus on to some general reflections about the teleological principle: "As regards the view that everything has a purpose and nothing is in vain, first of all the definition of this purpose is not so easy, as is often said; for where should we begin and where decide to stop? Moreover, it does not seem to be true of various things, some of which are due to chance and others to a certain necessity, as we see in the heavens and in many phenomena on earth" [175]. Theophrastus then proceeds to enumerate examples from the organic and inorganic world which are to all appearances without purpose and therefore cast doubt upon the existence of any tendency towards goodness and perfection.

However, the criticism of Theophrastus, devastating as it was despite the caution of its phrasing, so far from sweeping the teleological viewpoint out of existence, did not even succeed in noticeably disturbing it. In the end, Aristotle's cosmos continued to dominate human thought throughout the ancient and scholastic periods right up to the sixteenth century. Nor is this surprising, seeing that it was eminently qualified to maintain this position both in the Ancient World and after the rise of Christianity. In Aristotle's philosophy the cosmos is a sublime manifestation of the rule of order in the universe. This idea was as well suited to the Greek mentality, in which the concept of order blended with those of beauty and perfection as expressed in artistic creation, as to the basic creed of the monotheistic religions which regard the cosmic order as the work of the Creator and the expression of His will. The teleological idea therefore endured as a guiding principle in the explanation of nature, being woven into the pattern of mediaeval religious thought. Long life was likewise assured to the Aristotelian antithesis between heaven and earth, rooted as it was in star-worship, of which spiritual traces are to be found throughout the Greek period. In the theology of the monotheistic creeds this antithesis reappeared anew in the location of God and His angels in the heavens, in the pure region of Aristotle's eternal movements.

V

THE WORLD OF THE ATOM

"Here a little, there a little."
is. 28.10

THE science of Ancient Greece, in all its stages of development, shows the constantly recurring attempt to resolve the antithesis between the unity of the cosmos and the plurality of its phenomena. Is this antithesis real? Or is it perhaps only an illusion resulting from the imperfection of our senses—and if so, how does this come about? Is the plurality of phenomena only apparent, and the cosmos in fact a single unchanging unit, admitting of no movement? Or is it this unity that is imaginary, and reality in fact no more than the sum total of unending mutations and changes?

The answers given to these questions by the early Greek philosophers range from one extreme to the other. Thales and his followers held a monistic view, deriving everything from primeval matter; and Parmenides of Elea went so far in this monism as to rule out change and all possibility of movement. At the other extreme, we have seen in Empedocles the beginning of a pluralistic conception in his theory of the four elements; and Anaxagoras carried pluralism to its logical conclusion in his conjecture that while matter is continuous, it actually consists of minute quantities of all the various differentiations of which it is capable: in other words, the plurality of cosmic phenomena results from the plurality of qualities contained in every single particle of matter. Of all the answers, that given by the Greek

105

atomists, being a synthesis of monistic and pluralistic elements, is the most convincing both in its simplicity and in its comprehensiveness. It showed that the plurality and flux of the macrocosm can be explained by a certain uniformity and by causal laws governing the world not accessible to our senses.

The whole subject of the ancient atomic hypothesis is particularly instructive for us who live in the age of the modern atomic theory. The atomic structure of matter has now been proved by experiment and is for us a firmly established and unquestionable fact. The history of the past one hundred and fifty years shows us how this certainty only slowly emerged from general conjectures, the reality of which was regarded with scepticism by many scientists, and from a long series of interrelated experiments and deductions which eventually grew into a uniform and complete theory. In the light of this modern development, it is interesting to enquire how far the Greeks succeeded at a time when systematic experimentation was virtually unknown, when the methods employed consisted mainly of observation and speculative conjecture, and when the scientific principles and inferences in use were few in number and of the most general nature. There is no question here of comparing the two theories in terms of absolute scientific achievement: such a comparison would obviously be both pointless and unfair. The main purpose of the comparison is rather to estimate the validity of a method as shown in its internal logic and the extent to which it succeeded in developing to the full its basic premises at a time when scientific evidence was in the main qualitative. For we must bear in mind that the Ancient Greeks hardly knew of experimentation and mathematical deduction as means of applying scientific intuition to reality: for them analogy and the scientific model were the only connecting-links between the invisible and the visible.

The history of the ancient atomic theory extends over four hundred years and is connected with four famous names. Its author was Leucippus of Miletus, who lived in about the middle of the fifth century B.C. His pupil was Democritus of Abdera (born *c.* 460 B.C.), one of the most universal thinkers of the Ancient World. The theory was turned into a philosophical system by Epicurus of Samos (341-270 B.C.), who thus made it generally known. Finally, we must mention the Roman poet

T. Lucretius Carus (died *c.* 55 B.C.), whose didactic poem *De Rerum Natura* is a paean of praise in honour of the Epicurean philosophy. This poem is also the most detailed source of our knowledge about the atomic theory of Epicurus, apart from Epicurus' letter to his pupil Herodotus. Of the numerous writings of Democritus, only a few fragments are extant, and still less of Leucippus; but quite a number of their sayings are quoted by Aristotle and Theophrastus in the course of their strong polemic against the atomic theory. These quotations, together with some from Plutarch and the doxographic literature, complete our picture of the theory evolved by the founders of the atomic school. Some details in which Leucippus, Democritus and Epicurus differ from each other merit special attention as indications of the theoretical development within a particular scientific doctrine. But, in general, it may be said that the principles of the atomic theory are the same for all its proponents.

All these thinkers frequently emphasize the view that there is in nature a general conformity to law, and in particular that there exists a law which may be called "the law of the conservation of matter". Leucippus is quoted as saying that "nothing happens at random; everything happens out of reason and by necessity" [70]; and the atomic doctrine of Democritus contains the sentence: "Nothing can be created out of nothing, nor can it be destroyed and returned to nothing" [92]. This emphasis on the conservation of what exists is important for the proof of the existence of the atoms, as opposed to the theory that matter is infinitely divisible. It is known that Leucippus was once a pupil of the Eleatic School, which included Zeno of Elea whose famous paradoxes taxed the ingenuity of the best Greek philosophers. Some of these paradoxes, which we shall return to below, are based on the principle of dichotomy or halving. The argument runs that there is no end to the division of a section into parts, since there is no limit to any form of mathematical division, including halving. Thus the number of points between two given points is infinite. It is very probable that one of Leucippus' reasons for trying to find a solution to the problem of permanency in plurality was the paradox of division. This led him to the conclusion that physical division is not the same as mathematical division: "The atomists hold that splitting stops when it reaches indivisible particles and does not go on infinitely" [93]. This assumption of a lower limit to the division of matter is an axiom which may be proved by arguments of plausibility, as follows: if matter can be infinitely divided, it is also subject to complete disintegration from which it can never be put together again. In other words, if we wish to maintain the law of the conservation of matter and to consider the process of its disintegration and reintegration as a reversible one, we must assume that the disintegration or fragmentation stops at a definite and finite limit. Only so does there remain a permanent primary foundation for a new building up from the ultimate particles without any loss in the quantity of matter. Hence, in his letter to Herodotus, Epicurus says: "Therefore, we must not only do away with division into smaller and smaller parts to infinity, in order that we may not make all things weak, and so in the composition of aggregate bodies be compelled to crush and squander the things that exist into the non-existent . . ." [180].

As a corollary to this insistence on the existence of the atoms, Epicurus also insists on an upper limit to their size: they are always invisible, and a visible atom is inconceivable. Though no specific reason is given for this assumption, its empirical basis is obvious enough: bodies that can be seen are still divisible and therefore cannot be atoms. In contrast to Democritus, who did not postulate a maximum size for the atom, Epicurus apparently evolved the theory that the perceptible and imperceptible are two essentially different categories of existence.

The atomic theory was completed by a second axiom, the postulation of a "vacuum". The vacuum was introduced with rigorous consistency into the picture of the cosmos as an independent reality. Once again, this assumption is based on plausibility: given that matter is composed of atoms, of ultimate unchanging particles, then all changes must be the result of their movements, and the prerequisite of movement is a vacuum, a space entirely devoid of matter in which a particle can pass from place to place: "Unless there is a void with a separate being of its own, 'what is' cannot be moved—nor again can it be 'many', since there is nothing to keep things apart" [71]. On this view it necessarily follows that there is no possibility of a vacuum inside the atom itself, since in such a case the atom would be subject to changes and to physical influence from outside and would thus be likely to disintegrate. Hence the postulation of a vacuum as a

prerequisite for the movements of the atoms inescapably leads
to the postulation of the absolute solidity of the atom itself.
Matter and the vacuum are entirely separate from each other.
In the words of Aristotle's summing up: "Leucippus, however,
thought he had a theory which harmonized with sense-perception
and would not abolish either coming-to-be and passing-away or
motion and the multiplicity of things. He made these concessions
to the facts of perception: on the other hand, he conceded to the
Monists that there could be no motion without a void. The result
is a theory which states as follows: 'The void is a "not-being",
and no part of "what is" is a "not-being"; for what "is" in the
strict sense of the term is an absolute plenum. This plenum,
however, is not "one": on the contrary, it is a "many" infinite in
number and invisible owing to the minuteness of their bulk. The
"many" move in the void (for there is a void): and by coming
together they produce "coming-to-be", while by separating they
produce "passing-away"'" [72].

All the atomists, from Leucippus to Epicurus and his disciples,
are agreed that both the number of atoms and the extension of
the cosmos are infinite. The infinity of the cosmos in time, i.e.
its eternity, was deduced by Democritus from the conservation
of matter which rules out a *creatio ex nihilo*: "Democritus of
Abdera said that there is no end to the universe, since it was not
created by any outside power" [94]. On the other hand, infinity
of space and of the amount of matter in it are clearly interrelated.
It would appear from this that the essential point in both
premises is the assumption that space is infinite. Epicurus is quite
explicit about this: "Moreover, the universe is boundless. For that
which is bounded has an extreme point: and the extreme point
is seen against something else. So that as it has no extreme point,
it has no limit; and as it has no limit, it must be boundless and
not bounded. Furthermore, the infinite is boundless both in the
number of the bodies and in the extent of the void. For if on the
one hand the void were boundless, and the bodies limited in
number, the bodies could not stay anywhere, but would be
carried about and scattered through the infinite void, not having
other bodies to support them and keep them in place by means of
collisions. But if, on the other hand, the void were limited, the
infinite bodies would not have room wherein to take their place"

[176]. Two points should be noticed in this passage. The infinity of the universe is here proved by a geometrical argument of the very kind that was rejected by the atomists in relation to the division of matter. In this case the geometrical argument may be regarded as valid, since the problem here concerns spatial qualities. Until the discovery in modern times of non-Euclidean spaces and their properties, it was reasonable to reject the conception of finite space on grounds such as those advanced by Epicurus. It is likewise worth mentioning that Epicurus' argument to prove the impossibility of a concentration of a finite amount of matter within infinite space has also been used in the modern cosmological controversy.

The universe of the atomic school thus consisted of a vacuum of infinite size filled with "solid particles", atoms, of infinite number. In what do these atoms differ from each other? Certainly not in the matter which fills them, for this would contradict the monistic basis of the atomic theory. No; the primeval matter of which the atoms are made is uniform, but there are various kinds of atoms distinguished by their shapes: "The atoms are differentiated by their shapes: the nature of them all is, they say, the same, just as if, e.g. each one separately were a piece of gold" [73]. We are also told that "Democritus and Leucippus say that there are indivisible bodies, infinite both in number and in the varieties of their shapes, of which everything else is composed— the compounds differing one from another according to the shapes, positions, and groupings of their constituents" [74]. The physical, concrete qualities of macroscopic bodies are therefore determined by the particular kind, or combination of kinds, of their component atoms, and also by various principles controlling the dispositions of the atoms in the space occupied by the body, defined as "positions" and "groupings".

The shape of the atom corresponds to the chemical element of the modern atomic theory. It is the shape which differentiates the atoms in the same sense as the atomic number does to-day, determining their chemical properties. Since the ancient atomic theory was built upon purely mechanical conceptions, it is not surprising that the distinguishing marks of the atoms were mechanical or geometrical: "They have all sorts of shapes and appearances and different sizes. . . . Some are rough, some hook-shaped, some concave, some convex and some have other innumerable variations" [95]. According to the passages quoted above, Leucippus and Democritus maintained that the number of shapes was infinite, whereas Epicurus held it to be finite. This difference follows naturally from a variance in the assumptions about the size of the atoms. To every given size it is possible to assign only a finite number of distinctly different shapes, and once all the possible mutations have been exhausted, a fresh shape can only come into being through an increase in the volume of the atom. Hence the first atomists, who set no upper limit to the size of the atoms, did not restrict the number of their shapes either. But Epicurus was obliged by his proviso that the size of the atom must be invisible to reduce the shapes to a finite number: "Besides this the indivisible and solid bodies, out of which too the compounds are created and into which they are dissolved, have an incomprehensible number of varieties in shape; for it is not possible that such great varieties of things should arise from the same atomic shapes, if they are limited in number. And so in each shape the atoms are quite infinite in number, but their differences of shape are not quite infinite, but only incomprehensible in number" [176]. Lucretius, who repeats this sentence about the finitude of the different shapes of the atom almost word for word, gives an extremely vivid illustration of how the increasing number of shapes is linked with the increasing size of the atoms.

Leucippus apparently distinguished only the size and shape of the atoms. In Democritus, on the other hand, there is mention of weight, not however as an independent quality but as a function of the volume of the atom: "Democritus says 'the more any indivisible exceeds, the heavier it is'" [96]. "Exceeds" refers here to the volume of the atom, its extension in space. In this point there is a fundamental change in Epicurus' doctrine: "Democritus recognized only two basic properties of the atom: size and shape. But Epicurus added weight as a third. For, according to him, the bodies move by necessity through the force of weight" [97]. The last sentence shows that Epicurus found it necessary to introduce weight as the cause of the movement of the atoms; or, more precisely, as the reason for their falling, since in the cosmology of Epicurus "up" and "down" appear as absolute directions—an innovation no doubt to be traced to Aristotle's influence. Democritus, on the contrary, denied that

111

movement was due to weight, as is clear from the following passage: "Democritus said that the atoms have no weight, but they move by mutual impact in infinite space" [98]. The words "have no weight" are to be interpreted as meaning that weight is not the cause of movement, as is clear from the following sentence from Cicero: "The motive force that they will get from Democritus is a different one, a driving force termed by him a 'blow'; from you, Epicurus, they will get the force of gravity or weight" [99]. The picture drawn by Democritus reminds us of the atoms of the ideal gas in the modern kinetic theory of gases, which are kept in perpetual motion characterized by constant collisions.

Throughout the literature of the atomic school, great stress is laid on the perpetual movement of the atoms. But what is the origin of this movement? Aristotle severely criticizes the absence of a cause of the movements of the atoms in the doctrine of Leucippus and Democritus. In his book *On the Heavens* he writes: "When therefore Leucippus and Democritus speak of the primary bodies as always moving in the infinite void, they ought to say with what motion they move and what is their natural motion. Each of the atoms may be forcibly moved by another, but each one must have some natural motion also, from which the enforced motion diverges" [162]. The same question is raised again in his *Metaphysics*: "They say there is always movement. But why and what this movement is they do not say, nor, if the world moves in this way or that, do they tell us the cause of its doing so. Now nothing is moved at random, but there must always be something present to move it; e.g. as a matter of fact a thing moves in one way by nature, and in another by force or through the influence of reason or something else" [170]. It was a sound scientific instinct that saved the founders of the atomic school from this mesh of reasoning in which Aristotle got himself entangled. They did not begin by raising the problem of the cause of movement, but accepted movement as a given fact, just as they did in the case of the atoms. It is not wise to raise all the questions involved in a scientific problem simultaneously; on the contrary, a developed scientific sense is required to limit the range of questions at the start and to consider only some of the phenomena as derivatives of others while regarding the rest as ultimate data. Even without knowing the law of the conservation of momentum (quantity of movement), Leucippus and Democritus hit the mark when they assigned to every atom a predetermined movement and described the sum total of atomic movements by the elementary mechanical model of elastic collision. In this way they succeeded in grasping the first principles of the kinetic law of matter entirely without the aid of mathematics and with only the most primitive statistical concepts: "These atoms, which are separated from each other in the infinite void and distinguished from each other in shape, size, position and arrangement, move in the void, overtake each other and collide. Some of them rebound in random directions, while others interlock because of the symmetry of their shapes, sizes, positions and arrangements, and remain together. This was how compound bodies were begun" [75].

This remarkably clear passage leaves no room for doubt that, in the opinion of the atomists, not all the atoms form part of compound bodies, but there are always some left moving freely. Lucretius, in the second book of his poem, describes this phenomenon in great detail, emphasizing that atoms sometimes escape from the compound bodies and resume their free movement in the void. In the case of the atoms which interlock to form compound bodies, the atomists consistently maintained that they too continue in perpetual motion. Each of them continues to move in the narrow space bounded by its neighbours, being subject to very frequent collisions which are like swift vibrations in its narrow enclosure. This means that the interlocking of the atoms does not turn them into a single physical unit: even after combining each one maintains its individual character, as shown in its movements, which in these circumstances take the form of vibrations.

In this connection, the two following passages from Aristotle and his commentator, Simplicius, are most instructive: "According to this view the primary magnitudes are infinite in number and not divisible in magnitude. Generation is neither of many out of one, nor of one out of many, but consists entirely in the combination and entanglement of these bodies. In a way these thinkers too are saying that everything that exists is numbers, or evolved from numbers" [76]. This last sentence of Aristotle's, which was intended as a refutation, from our modern standpoint actually throws into relief the great advantage of the atomic hypothesis—that it gives quantitative values to phenomena.

Simplicius, too, emphasized the primary identity of every atom: "As a result of their movement they strike each other and get caught in an entanglement which brings them in contact with each other and makes them come very close together. But any real unity is not formed out of them. That would be an utterly foolish opinion, since two or many things can never become one. . . . In his opinion, they hold out and remain together until some stronger force acts upon them from outside, shaking them and scattering them" [95].

In his detailed discussion of the particles composed of many atoms, Epicurus finds it difficult to explain the transition from atomic movement to the movement of large bodies. According to his theory all the atoms have the same, extremely great, velocity. How, then, can the slower velocity of a compound body moving in a certain direction result from a combination of atoms moving at an enormous speed in all directions? Unfortunately, the relevant parts of Epicurus' letter to Herodotus are corrupt; but it would seem that the obscurities cannot be attributed only to the corruption of the text. Without going into the details, it is worth mentioning this case as an outstanding example of a characteristic phenomenon in the scientific literature of Ancient Greece: sometimes thinkers, whose penetrating scientific imagination and profound grasp of complex facts truly amaze us, are at the same time perplexed by questions which now seem relatively simple to anyone acquainted with the elementary laws of kinetics and mechanics. One aspect of this paradox deserves to be examined further. While it is true that the basic concepts of mechanics were only properly worked out in the seventeenth century, it still cannot be denied that serious, if unsuccessful, attempts were made in the Ancient World to clarify the interrelation of velocity, force and mass. We have seen how Aristotle himself made considerable efforts to formulate the laws of dynamics. But, for all that, the ancients had no conception of statistics, for determining the laws of phenomena where very large numbers of individuals or very large numbers of repetitions of a given occurrence are involved. We do not find amongst them anything resembling the "law of large numbers", or the law of averages, or the like, even though the games of chance which were so common in the Ancient World provided plenty of opportunities for their study. (This question will be dealt with at length in a later chapter.)

The one exception to this rule occurs, as might be expected, in the atomic theory, which deals with an enormous number of individuals. On this point we possess two famous descriptions of parts of Epicurus' doctrine, both of them in the second book of *De Rerum Natura*. In the first, Lucretius discusses the case of many particles moving in all directions within given boundaries. If this is so, the sum total of these particles will be at rest as a single entity in the given space; or, in other words, the total of all the velocities (if we add them together as vectors, i.e. we take their directions into account) will be zero. This, as every student of elementary physics knows, is how we picture the atoms of a gas confined within a certain volume. Lucretius, who of course did not know this law, attributes the apparent rest to the inability of our senses to discern the moving atoms. Whether his reasoning was right or wrong does not matter here. The essential thing is that he grasped the fact that every group made up of particles moving at random will appear to be a single body at rest. These are his words: "In this connection there is one fact that need occasion no surprise. Although all the atoms are in motion, their totality appears to stand totally motionless. . . . This is because the atoms all lie far below the range of our senses. Since they are themselves invisible their movements also must elude observation. Indeed, even visible objects, when set at a distance, often disguise their movements. Often on a hillside fleecy sheep, as they crop their lush pasture, creep slowly onward, lured this way or that by grass that sparkles with fresh dew, while the full-fed lambs gaily frisk and butt. And yet, when we gaze from a distance, we see only a blurb—a white patch stationary on the green hillside" [249].

The second description concerns a much more complex phenomenon, one which was discovered by the microscope in the first half of the nineteenth century but was not fully explained and reduced to mathematical terms until the beginning of the twentieth century. This is the Brownian movement. When we look at microscopic particles suspended in a liquid or a gas, such as oil-drops in an emulsion or dust and smoke particles in the air, we see how they move in a perfectly disorderly fashion, wandering this way and that without rule and without purpose. This indirectly shows us the activity of the atoms in the liquid or the gas which cannot be seen even in a microscope. It is true that in the average taken over a longer period of time the total of all the impacts of the atoms on the microscopic particles is cancelled out. But the statistical deviations from the average occurring at every moment result in the particle's being incessantly given impulses this way and that with constant changes of direction, and it is random impulse which gives rise to the ceaseless oscillations of the particles. This is the phenomenon that we find described and concretely illustrated in a passage from Lucretius which is remarkable for its clarity and the way in which it picks out the main points: "This process, as I might point out, is illustrated by an image of it that is continually taking place before our very eyes. Observe what happens when sunbeams are admitted into a building and shed light on its shadowy places. You will see a multitude of tiny particles mingling in a multitude of ways in the empty space within the light of the beam, as though contending in everlasting conflict, rushing into battle rank upon rank with never a moment's pause in a rapid sequence of unions and disunions. From this you may picture what it is for the atoms to be perpetually tossed about in the illimitable void. To some extent a small thing may afford an illustration and an imperfect image of great things. Besides, there is a further reason why you should give your mind to these particles that are seen dancing in a sunbeam: their dancing is an actual imitation of underlying movements of matter that are hidden from our sight. There you will see many particles under the impact of invisible blows changing their course and driven back upon their tracks, this way and that, in all directions. You must understand that they all derive this restlessness from the atoms. It originates with the atoms, which move of themselves. Then those small compound bodies that are least removed from the impetus of the atoms are set in motion by the impact of their invisible blows and in turn cannon against slightly larger bodies. So the movement mounts up from the atoms and gradually emerges to the level of our senses, so that those bodies are in motion that we see in sunbeams, moved by blows that remain invisible" [247].

To this remarkable description we need only add the comment that it perfectly describes and explains the Brownian movement by a wrong example. The movements of dust particles as seen by the naked eye in sunlight are caused by air-currents; the real phenomenon postulated by Lucretius on the basis of abstract reasoning can only be seen in a microscope. However, this stricture in no way detracts from the importance of the discovery itself. It may be said that the greatest achievement of the atomic school in Ancient Greece was the introduction into scientific reasoning of the method of inference, as demonstrated in this passage of Lucretius. It is only natural that the atomic theory should have awakened scientific thought to the possibility of inference from the visible to the invisible. In his first book, Lucretius had already given proofs of the limitation of our senses which conceals from us details of processes whose reality cannot be doubted, since we eventually become aware of them: "I have taught you that things cannot be created out of nothing nor, once born, be summoned back to nothing. Perhaps, however, you are becoming mistrustful of my words, because these atoms of mine are not visible to the eye. Consider, therefore, this further evidence of bodies whose existence you must acknowledge though they cannot be seen. . . . We smell the various scents of things though we never see them approaching our nostrils. Similarly, heat and cold cannot be detected by our eyes, and we do not see sounds. Yet all these must be composed of bodies, since they are able to impinge upon our senses. For nothing can touch or be touched except body. Again, clothes hung out on a surf-beaten shore grow moist. Spread in the sun they grow dry. But we do not see how the moisture has soaked into them, nor again how it has been dispelled by the heat. It follows that the moisture is split up into minute parts which the eye cannot possibly see. Again, in the course of many annual revolutions of the sun a ring is worn thin next to the finger with continual rubbing. Dripping water hollows a stone. A curved ploughshare, iron though it is, dwindles imperceptibly in the furrow. We see the cobble-stones of the highway worn by the feet of many wayfarers. The bronze statues by the city gates show their right hands worn thin by the touch of travellers who have greeted them in passing. We see that all these are being diminished, since they are worn away. But to perceive what particles drop off at any particular time is a power grudged to us by our ungenerous sense of sight" [245].

To appreciate this and similar passages at their true worth, we must remember how important a function in the explanation of

a phenomenon, or the interpretation of an experiment, is fulfilled
by this kind of scientific reasoning even in our days of experimental science and mathematical formulation. This essential
element of the scientific method reached a peak of development
in the Greek period. In this the atomic school undoubtedly played
a decisive part, though examples could be adduced from other
schools as well.

We have seen that the basic premisses from which Leucippus
and Democritus started were the existence of a vacuum and of
atoms differentiated by shape, position and arrangement. It is
now natural to enquire to what extent these thinkers tried to
infer all the consequences of these premisses and to build upon
them a physical or chemical theory of matter as a rational
explanation of physical phenomena. The answer to this question
touches upon the philosophical background of the atomists. Their
approach to all natural problems, including both biological and
psychological phenomena, was rigorously mechanistic: they saw
everything as due to the movements of matter and to the contacts
between its parts, starting from the creation of the universe and
ending with man's senses and soul. For them, there was no
question of introducing any other motive force as the cause of
physical processes: such forces, being "irrational", must
eventually break up the mechanistic picture of the universe by
becoming firmly entrenched in it under the guise of "spiritual"
causes, like the Mind of Anaxagoras; or they may even lead to
the confusion of entirely different categories of existence by
turning the gods into the supreme cause. By thus compre-
hensively ruling out the existence of forces, Democritus and his
followers left themselves with only one cause for the explanation
of all physical change—impact, or the collision of atoms or
aggregates of atoms.

The application of this principle to epistemology led the
founders of the atomic school to take up the same position as
Locke and the English Empiricists in the eighteenth century. The
objective basis of sensation is simply and solely contact. This is of
two kinds: either direct contact between the person perceiving
and the object perceived, as in touching, or tasting; or contact
between the person and the atoms emitted from the object and
entering his nose, ear or eye. Thus Democritus, like Locke,

distinguishes the "secondary qualities" of bodies—colour, smell, taste, sound—which are the subjective product of our senses and can be explained by the mechanical attributes of the atoms, from their "primary qualities", such as impenetrability, hardness, etc., which are the objective expressions of the "true" attributes of matter. Democritus' views on this point have been preserved verbatim by Galen: "Colour exists by convention, sweet by convention, bitter by convention; in truth nothing exists but the atoms and the void" [100]. To this Galen adds: "This is what Democritus says, and also that all the qualities of the things perceived by us result from the collision of the atoms. In reality there is no white, or black, or yellow, or red, or bitter, or sweet. For what he calls 'by convention' means 'according to usage', or 'from our standpoint', and not 'as things really are' which is what he calls 'in truth'." Again: "Others said that the perceived qualities are of the nature of things, but Leucippus, Democritus and Diogenes said that they are essentially matters of convention, i.e. they come from our mind and our impressions" [77].

Since the writings of Democritus are not extant, we have to rely on the references of other writers for our knowledge of his theory of matter. All in all, these amount to a feeble enough echo which is particularly difficult to interpret just on the details of most interest to us. That many passages became obscure in the course of transmission is made probable by various contradictions and by the fragmentary and unintelligible condition of several texts. A considerable part of the whole theory has been preserved in the writings of Theophrastus, particularly in his book *On Perception and Things Perceived*. But here, too, the detail and clarity of the exposition are very uneven. Still, there can be no question that Democritus tried to explain both the primary and secondary qualities, i.e. all the macroscopic properties of things, by the attributes inhering in the primary elements of nature. In this he was followed by Epicurus. Once again a few examples will reveal a characteristic picture: although the theory is purely speculative and contains hardly a single detail capable of withstanding the criticism of modern science, the ways of thinking of the ancient atomists were, none the less, essentially the same as those of our atomic theorists. Both alike aim at explaining the sensory qualities of objects by physical changes in the realm of the atom.

What explanation of the weight of a macroscopic body was given by the atomists? As we have seen, specific weight, i.e. the weight of a body's unit of volume, was unknown to the ancients before the time of Archimedes (287-212 B.C.). Archimedes not only had a clear comprehension of this concept, but also discovered how to determine the relative density of bodies by experiment. Before his time, complete confusion had reigned with regard to this elementary quantity, a confusion which had reached its climax in Aristotle's theory that weight and lightness were absolutes. To Democritus it was clear that the weight of a body depended on the admixture of fullness and emptiness in it, i.e. on the atoms and the distance between them. Translating this into modern scientific terms, we may say that he understood that the weight of the body depends on its lattice structure and the atomic weight. However, the words of Theophrastus are ambiguous: "Democritus distinguishes heavy from light by the size (of the atoms). For if any object is decomposed into its parts, the weight of its elements will be in accordance with their size, even if they are different from each other in shape. But in compound bodies, that which contains more of the void is light while that which contains less is heavy" [101]. Let us give these words a liberal interpretation and suppose that Democritus realized that weight is a function of both the heaviness of the kind of atom (or mixture of several kinds) of which the body is composed and also of the number of atoms in a given volume. The problem would then be solved "in principle", although in practice this realization would be of no value at all, since Democritus knew of no way of measuring these two quantities. Even so, his approach to this problem is itself worthy of note. Equally interesting is his attempt to explain the differences in another primary quality, the hardness of bodies: "He explains hard and soft in a similar way. Hard is what is dense, and soft what is rare. . . . Hard and soft as well as heavy and light are differentiated by the position and arrangement of the voids. Therefore iron is harder and lead heavier. For iron is not regularly compounded, but contains large voids in many places, while in other places it is dense, the void being on the whole predominant. Whereas lead contains less void and is regularly compounded and evenly on the whole. Therefore, although it is heavier than iron, it is at the same time softer than it" [101].

It is not easy, from this passage, to obtain a clear picture of Democritus' "theory of the solid state" or "crystallography". He apparently supposes that an atom of iron and an atom of lead have the same weight and that the heaviness of the materials is therefore determined by the number of atoms in a given volume, those of lead being more numerous than those of iron. But,

FIG. 7. The lattice structure of lead (1) and
iron (2) according to Democritus.

whereas the lead atoms are arranged regularly, the iron atoms are irregularly disposed in space, with the result that a number of them are closer together than the regular distance between the lead atoms. Although found in only a small portion of the bulk of iron, it is this great density that determines the greater hardness of this substance. Perhaps this was Democritus' way of making the principle of atomic density account for two such different qualities as heaviness and hardness, which are not always found together in the same substance. But it is not impossible that Democritus—for some reason unknown to us—attributed the small strength and resistance of a substance to its symmetrical atomic structure, as in the case of the softness of lead: in another passage from Theophrastus the brittleness of certain bodies is explained by supposing that their atoms are arranged "evenly".

The description of the origin of the secondary qualities is more detailed, especially for colours and tastes. Here there is a noticeable and significant shift of emphasis between Democritus and Epicurus, which is of interest if understood as an advance in scientific thought. In theory, Democritus attaches equal importance to the shape, position and arrangement of the atoms as the cause of sensory phenomena. But in practice, when he comes to explain colour and the other secondary qualities, he regards shape as the main cause. As against this, Epicurus restores position and arrangement to their place of equal importance here too, thus taking the first step from an atomic theory to a molecular one. First let us hear Democritus, who distinguishes four basic colours:

121

white, black, red and green: "White is made up of smooth atoms, for what is not rough, nor obscuring nor hard to penetrate is wholly bright. For it is necessary that bright bodies should have straight pores and should be translucent. . . . Black is made up of atoms of the opposite kind, jagged, with unequal sides and dissimilar. Hence their pores are dark and not straight and are not easily penetrated by light. . . . Red is made up of the same atoms as hot, but larger. . . . An indication that the red atoms are of the same kind as hot ones is the fact that we get red when heated and all burning bodies are red as long as they contain fiery matter. . . . The simple colours are composed of all these atoms: the smaller the extent of the mixture of its atoms with others, the purer each one is. The other colours are a mixture of all these" [102].

In this connection mention must be made of the atomists' theory of vision which was derived from Empedocles' theory of emanations. According to this, emanations are emitted from objects and enter the eye, an idea which was also borrowed for the explanation of magnetic phenomena. As regards the theory of vision, we are told that "Leucippus, Democritus and Epicurus say that vision is essentially the entry of images" [78]. Another source goes into more detail: " Anticipated by Leucippus and followed by Epicurus and his disciples, Democritus said that a sort of images emanate from bodies, that they are of the same shapes as the bodies from which they emanate and strike the eye of the beholder and so vision is produced " [79].

This theory of images underlies the whole atomistic doctrine of sensation. According to Democritus, atoms of taste enter a man's body through his mouth, while atoms of smell come from the objects to his nose, where their effect is essentially determined by their shape: "Sharpness is composed of angular and many-cornered atoms which are small and fine, since sharpness quickly spreads everywhere, and being rough and angular they produce contraction. Hence the body grows warm, since empty spaces are formed in it. For the larger the void in the body, the hotter it becomes. Sweetness is composed of round atoms which are not small. It therefore spreads throughout the body and penetrates everywhere gently and slowly. . . . There are no pure kinds of atoms which are free from admixture. Every taste contains many kinds, being composed of smooth, rough, round, angular atoms, etc. The larger the proportion of any one kind, the larger its share in the sensation and its strength" [103].

Perception varies noticeably with the perceiver, especially in the case of taste and smell. In order to fit this fact into his mechanistic theory, Democritus adds: "Not a little depends on the state of the body which they enter. For this reason, one kind of atoms sometimes produces the opposite sensation, while the opposite produces the original one." If we accept the evidence of Theophrastus, we may deduce that in all cases Democritus supposed that in fact the shape of the atoms, and not their combination, was decisive. Even if this was not actually his view, such at least was the impression which he succeeded in giving his commentators. This is particularly striking in the explanation of phenomena such as qualitative changes occurring in a substance over a period of time; for example, the changes in the taste of juices exuded from various plants which are e.g. bitter at first, but subsequently become sweet, and so on. In his book *On the Causes of Plants* Theophrastus writes: "It is not clear how Democritus would have us conceive the creation of different kinds of juices from one another. For there are three possibilities: either the atoms change their shape from rough and angular to round; or all the shapes of atoms are contained together in sour and sharp and sweet, some of them being separated out while others remain; or, finally, some go out while others enter. Now, since it is impossible for the atoms to change their shape, the atom being impassive, there remain two possibilities: either some enter while others go out, or some remain, while others depart" [104].

Difficulties of this kind in explaining changes in qualities were removed by the important transition from the atomic theory to the molecular one, the credit for which belongs to Epicurus. Illustrations of this new conception are found in the poem of Lucretius. How is it possible to explain sudden changes, such as occur, for example, in the colour of the sea from dark one moment to bright the next, or in the colour of various objects? The answer must be that a given colour results from a given combination of atoms and any change in this combination produces a change of colour. There is thus no need to postulate a mixture of various kinds of atoms in the object, some of which are emitted while a new kind is added from outside. Here the shapes of the atoms are still mentioned, but the main emphasis has been shifted to their combinations: "Let us suppose, then, that the atoms are naturally colourless and that it is through the variety of their shapes that they produce the whole range of colours, a great deal depending on their combinations and positions and their reciprocal motions. You will not find it easy to explain without more ado why things that were dark-coloured a moment since can suddenly become as white as marble. . . . You could say that something we often see as dark is promptly transformed through the churning up of its matter and a reshuffling of atoms, with some additions and subtractions, so that it is seen as bleached and white" [250]. Lucretius also emphasizes the part played by light in the creation of colour. As proof of this he mentions the obliteration of all colours by darkness and the change of colour produced by changes in the angle of incidence "according as to whether the light rays strike the body directly or at an angle".

Epicurus himself, in his letter to Herodotus, also speaks of "something . . . which can cause changes; not changes into the non-existent or from the non-existent, but changes affected by the shifting of position of some particles, and by the addition or departure of others" [179]. And Lucretius, when summing up the characteristics of the secondary qualities, reiterates: "So you may realize what a difference it makes in what combinations and positions the same elements occur, and what motions they mutually pass on and take over" [251]. Both here and in the earlier passage [250] these atomic combinations are described as having movements which are co-ordinated in a certain way, an idea which is characteristic of the whole Epicurean conception of the nature of the molecule. We must remember that Epicurus did not admit the existence of forces which hold the atoms of the molecule together thus making them into a single entity, though he almost certainly postulated a mechanical cohesion resulting from the jagged, easily interlocking shapes of the atoms.

However, according to Epicurus, the molecule has another distinguishing mark, namely the co-ordination of the movements of atoms composing it—the movements which these atoms "mutually pass on and take over" [251]. We have already seen that the perpetuity of motion was one of the basic premises of the atomic school, being applied even to the "compound bodies"

[75] in which the atoms are so close together that the motion inside them takes the form of a vibration resulting from the rapidly repeated collisions and recoils. The simplest compound body of all is the molecule, called by Lucretius "concilium", which means union or association and is close to our modern concept of a chemical compound. The compound is a unit of a higher order than the atom, and its structure is closely associated with the nature of the motion of its components. In one passage of Lucretius' poem this fact is stated in unmistakable terms. When Lucretius describes the movement of the atoms and the creation of more complex bodies by their collisions, he says: "Besides these, there are many other atoms at large in empty space which have been thrown out of compound bodies and have nowhere even been granted admittance so as to bring their motions into harmony" [246]. Co-ordination of the movements of the atoms in the molecule, harmony between their various vibrations, governed by a principle regulating their mutual movements—these are the physical factors which characterize the association of atoms, the concilium, and make it a single entity. Once again we are amazed at the imaginative power and scientific intuition displayed in the emphasis laid here on one of the characteristics of the molecule—namely, the sum total of its possible vibrations and their combinations, which Epicurus of course regards simply as a function of position and arrangement and not of forces. It may be that this model of the molecule was to some extent the result of observation. The ancients were aware that the mechanical movement of a body consisting of many parts held loosely together by chains or ropes depends on the form of these connections and that there is in such cases a kind of "communal motion" of all the parts in which the rhythm of each one is conditioned by the rhythm of the whole. The logical way of passing from this model to the picture of the molecule would have been to put forces in place of ropes as the cause of molecular vibrations. Only the Greek atomists, being opposed to the assumption of forces, regarded the co-ordinated motion within the association of atoms as the result of their mutual arrangement and the nature of their internal recoils, which, in their turn, were determined by the shapes of the individual atoms.

Since Epicurus considered the secondary qualities as originating mainly in the molecules, it is not difficult to understand his assumption that any change in the molecule, resulting from a change in the arrangement of its atoms, produces a change of colour or of taste and smell. It is clear why, on the basis of his mechanistic theory, he should ascribe mechanical causes to such changes of the molecules, causes such as moving or shaking which put an end to one kind of atomic association and bring others into existence. At the same time, Epicurus assumed that the molecules were structurally strong enough to continue existing as units even in isolation when they escape from bodies and pass from place to place. He took the theory of images which Democritus had used to explain the origin of sensation and applied it to the molecules: "Moreover, there are images like in shape to the solid bodies, far surpassing perceptible things in their subtlety of texture. For it is not impossible that such emanations should be formed in that which surrounds the object, nor that there should be opportunities for the formation of such hollow and thin frames, nor that there should be effluences which preserve the respective position and order which they had before in the solid bodies: these images we call idols" [177]. According to Epicurus, the speed at which these images travel is very great and they arouse sensations in us whenever they strike our bodies. There is no need to discuss the details of Epicurus' theory of sensation, with its very primitive mechanistic approach. But it is of interest to note the repeated emphatic assertion that the structure which the molecule has within the body—representing on a small scale all its properties—is preserved also after its emission; the molecule remains intact even at the moment of its impact upon our sense organs and imparts to them "the vibration of the atoms in the depth of the solid body" [178].

Undoubtedly it was Epicurus who actually evolved the molecular theory and tried to define the physical characteristics of molecules. To say this is in no way to detract from the achievement of Leucippus and Democritus, who were the first to conceive the molecular idea when they stressed the influence of the position and arrangement of atoms. Indeed, it would appear to be Democritus who was the author of an analogy which was intended to exemplify the nature of the molecule and which is very characteristic of the synthetic approach of the Greeks. The analogy is mentioned by Aristotle in his *Metaphysics*: "These philosophers say the differences in the elements are the causes of all other qualities. These differences, they say, are three—shape and order and position. For they say the real is differentiated only by 'rhythm' and 'inter-contact' and 'turning'; and of these rhythm is shape, inter-contact is order, and turning is position; for A differs from N in shape, AN from NA in order, ⊥ from H in position" [80]. First of all, the use of an analogy from language to explain a physical theory is in itself most instructive. The common point is, in this case, the construction of more complex units from units which cannot be broken down any further. This is also reflected in the Greek term "stoicheion" which is used both as a collective noun for "letter" and as one of the many Greek equivalents of "atom", and in a still more general sense of the ultimate elements of physical reality. Plato uses the word in this double sense in the *Theaetetus*, when explaining the difference between these elements, which cannot be analysed further, and their combinations. To prove his case he cites the analogous difference between letters as primary elements of language and the syllables constructed from them, which alone have any meaning. This comparison appears again in Plato's *Sophist*, this time with the usual word "gramma" for letter: there are some things in reality which combine with others, whereas in other cases such a combination is impossible. The letters of the alphabet, for example: some letters can form a harmonious combination, while there are others which cannot combine to form a more complex unit.

Democritus' picture is the aptest of all. In accordance with his view that the atoms are literally the primary units of matter, he regarded the atomic character of the letters of the alphabet as symbolical of the structure of the physical universe. The letter-atoms, which are devoid of the quality of meaning and are differentiated only by their shapes, combine to form syllables and words which are functions of position and arrangement. These combinations are linguistic "molecules", compounds which are units of a higher order with a definite meaning and which in turn combine to form the highest linguistic entity of all—sentences which are the vehicles of significant content. Just as a word is more than the algebraic sum of its component letters, so the particular association of atoms in a molecule is something different from the geometrical combination of elements. An entity is formed which, by force of its specific constitution, receives a specific quality which we perceive as colour, taste or smell. Any shifting of one of the constituting units from one place to another, any alteration in a single one of the elements, produces a complete change in the characteristic quality of the whole. In another passage Aristotle repeats this idea and adds: "For Tragedy and Comedy are both composed of the same letters" [81]. Aristotle, though strongly disagreeing with the atomic theory, was evidently most impressed by this picture of the raw material of the alphabet taking on such different shapes by virtue of structural changes. It is, therefore, not surprising that in the same passage he praises Democritus as against his predecessors: "Not one of them penetrated below the surface or made a thorough examination of a single one of the problems. Democritus, however, does seem not only to have thought carefully about all the problems, but also to be distinguished from the outset by his method" [82].

About three hundred years later Lucretius returned to the same analogy in the continuation of his words quoted above: "You will thus avoid the mistake of conceiving as permanent properties of the atoms the qualities that are seen floating on the surface of the things, coming into being from time to time and as suddenly perishing. Obviously it makes a great difference in these verses of mine in what context and order the letters are arranged. If not all, at least the greater part is alike. But differences in their position distinguish word from word. Just so with actual objects: when there is a change in the combination, motion, order, position or shapes of the component matter, there must be a corresponding change in the object composed" [251].

To sum up, then, it would seem that the development of the atomic theory from Leucippus and Democritus to Epicurus and his school is more a matter of progress in the clarification of details than of advance in scientific ideas or principles. An exception is the concept of the molecule, the association of atoms: here Epicurus made a striking original contribution to scientific knowledge. But, with this one exception, it may be said that the later writings in this field were merely explanations and commentaries of what had been discovered by the originators of the theory. The greatest achievement of the atomists was to develop a new kind of scientific reasoning based on evidence by analogy and inference of the invisible from the visible, by means of parallels and models as illustrations. We have seen that this achievement was no mere chance but a logical consequence of the main principles of the theory which set physical reality upon an infra-sensory basis. However, Democritus did not confine this scientific approach to the microcosm. We remember the analogy used by Lucretius when he explained the apparent continuity and immobility of matter by the model of a flock of sheep at pasture. Democritus had already used a similar analogy to explain the nature of the Milky Way: he describes it as "the confluence of rays of many, small continguous stars which shine together because of their density" [105]. According to another source, he said: ". . . that it is composed of very small, tightly packed stars which seem to us joined together because of the distance of the heavens from the earth, just as if many fine grains of salt had been poured out in one place" [106].

In this connection, however, regression, or what we might call a somewhat reactionary attitude can be discerned in the Epicurean philosophy of science. Epicurus had no room for a theory that in any way reintroduced divine activity into the realm of celestial phenomena. He held that once a supernatural cause were admitted, it would eventually end by destroying man's spiritual quietude by enslaving him to forces beyond his control. Hence the paradox that, in all that concerns astronomical phenomena, which are macroscopic, Epicurus insisted that the scientific explanation be attempted with a caution and suspension of judgment which he never thought of demanding for the invisible world of the atom. Here he was fully confident of the truth of his explanation. But "this is not so with the things above us: they admit of more than one cause of coming into being and more than one account of their nature which harmonizes with our sensations" [183]. Epicurus felt that at this very point it was better for him to abandon the scientific principle of a single consistent explanation than to be misled himself, or to mislead others, into believing that the heavenly phenomena are subject to law, which is tantamount to believing in the existence of the gods.

This point will be treated at greater length later. Here we shall conclude by mentioning another typical difference of conception between Democritus and Epicurus which again reveals Democritus as being more cautious and critical. We refer to their respective opinions about the nature of knowledge. Galen, in one of his writings, repeats the famous statement of Democritus with a very characteristic addition: "After Democritus had attacked sensation by saying that colour exists by convention, sweet by convention, bitter by convention, atoms and void exist in reality, he lets the senses say the following words against the mind: 'Miserable mind, you get your evidence from us and do you try to overthrow us? The overthrow will be your downfall'" [107]. Democritus was extremely doubtful about the value of the senses as tools of knowledge. Notwithstanding that his view of the world was absolutely rational, he realized that the mind has no choice but to use that most imperfect and unreliable of instruments, the senses. In another passage we are told: "There are two sorts of knowledge, one genuine one obscure. To the obscure belong all the following: sight, hearing, smell, taste, touch. The genuine is separated from this. . . . When the obscure can do no more—neither see more minutely, nor hear, nor smell, nor taste, nor perceive by touch—and a finer investigation is needed, then the genuine comes in as having a tool for distinguishing more finely" [89].

In Epicurus we find the opposite opinion. The philosophic scepticism of Democritus is replaced by a naïve realism and an unquestioning faith in the senses. To Epicurus' mind, any questioning of a single one of the senses is likely to remove the solid ground from beneath our feet: "If you fight against all sensations, you will have no standard by which to judge even those of them which you say are false" [187]. It is not the senses that represent a danger to our knowledge, but the deductions made by the mind from our sensation. Lucretius deals at length with this question: "You will find, in fact, that the concept of truth was originated by the senses and that the senses cannot be rebutted. The testimony that we must accept as more trustworthy is that which can spontaneously overcome falsehood with truth. What then are we to pronounce more trustworthy than the senses? Can reason derived from the deceitful senses be invoked to contradict them, when it is itself wholly derived from the senses? If they are not true, then reason in its entirety is equally false. . . . Whatever the senses may perceive at any time is all alike true" [252].

In view of the gulf which separates this opinion from the careful and balanced position taken up by Democritus, it is no wonder that, in the perspective of history, the later commentators linked together Democritus and Plato, one of them even comparing the 'Ideas' of the latter to the Atoms of the former: "The schools of Plato and of Democritus say that there is no reality except that comprehended by the mind. Democritus says so, because in his opinion the fundamentals of nature are not perceptible to the senses, seeing that the atoms of which everything is composed are by their nature devoid of all sensory attributes; while Plato says so, since to his mind things perceptible to the senses are in a constant process of being created and are not permanent" [108]. A few extant fragments of Democritus' writings confirm his critical stand, which was a logical outcome of his rationalism and a pessimistic view of the power of human knowledge: "We know nothing about anything really, but Opinion is for all individuals an inflowing" [85]. "It will be obvious that it is impossible to understand how in reality each thing is" [86]. "We know nothing accurately in reality, but as it changes according to the bodily conditions, and the constitution of those things that flow upon (the body) and impinge upon it" [87]. "One must learn by this rule that Man is severed from reality" [84]. The pessimism expressed here by one of the founders of the atomic theory reaches its climax in the following sentence: "We know nothing in reality; for truth lies in an abyss" [88].

VI

THE WORLD OF THE CONTINUUM

"And the breath came into them, and they lived."
EZ. 37.10

B EGINNING with the third century B.C., there grew up side
by side in Greece, and subsequently at Rome, two opposed
and rival physical theories, each of which became part of a
comprehensive philosophical doctrine. One, the atomic theory,
which is simply the teaching of Leucippus and Democritus with
extensions and modifications, was incorporated in the Epicurean
philosophy. The other is the original creation of the Stoics and
is associated principally with the names of Zeno of Cition in
Cyprus (c. 332-262 B.C.), Chrysippus of Soli in Cilicia (c. 280-207)
and Poseidonius of Apamea in Syria (c. 135-51). The corner-stone
of Stoic physics is the concept of a continuum in all its aspects—
space, matter and continuity in the propagation and sequence of
physical phenomena. The originality of the Stoic continuum
theory deserves to be emphasized, since Aristotle, too, in his
polemic against the atomic theory of Democritus, denies the
existence of a void, principally on grounds drawn from his own
definition of motion, and insists on the continuity of matter in
the universe. This insistence of Aristotle's, however, is no more
than an incidental aspect of his physical doctrine. For him the
continuum is essentially passive, whereas the Stoics transformed
it into an active quality and made it the governing principle in all
the physical phenomena of the cosmos.

The active substance in the cosmos which binds it firmly

132

together into a single dynamic whole is "pneuma", the Greek word for "spirit" or "breath". The term "pneuma" appeared in Greek physics as early as Anaximenes. At first it was usually no more than a synonym for air. Later, in the Stoic teaching it came to be used for a mixture of fire and air which embodied the characteristic quality of these two elements—viz. their activity— in a more pronounced form. According to Aristotle's definition, the quality common to the active elements is heat, fire being hot and dry, while air is hot and moist. The Stoics defined the elements differently, attributing only one quality to each, thus making fire hot and air cold, earth dry and water moist. But the Stoics agreed with Aristotle in attributing active qualities to fire and air. This they did for two main reasons, the first physical the second biological. The elastic qualities of air had long been known and its compressibility had been proved by experiments with air compressed into skins. In the Alexandrine period, notice also began to be taken of the expansive force of steam and similar phenomena indicative of vapour tension. But this physical reason was overshadowed by the position held by heat in biology. That organic life and the phenomenon of organic growth and biological development are inseparable from thermic processes was common knowledge to biologists and students of nature as early as the pre-Socratic period. Thus Cicero and others inform us that "Zeno gives this definition of nature: 'nature (he says) is a craftsmanlike fire, proceeding methodically to the work of generation'" [188]. But it was chiefly in the time of the later Stoics that a revolutionary advance was made: the dynamic functions of fire and air were extended to embrace all natural phenomena, including those of pure physics. From a certain standpoint this may be called a first tentative approach to the conception of thermodynamic processes in the inorganic world, a conception which began to percolate through into the scientific view of later generations. When this idea of the ubiquity of thermodynamic processes finally took root, it brought about a reversal of the original order, so that general thermic phenomena began to be used as proof and illustration of organic thermic processes. This we can see from a sentence of Cicero, perhaps quoted from his teacher Poseidonius: "All things capable of nurture and growth contain within them a supply of heat, without which their nurture and growth could not be possible; for everything of a hot, fiery nature supplies its own source of motion and activity; but that which is nourished and grows possesses a definite and uniform motion" [189].

The history of the concept "pneuma" as formed of fire and air is most instructive for the understanding of the Greek "biological" approach to the cosmos. The Greek projected the functions of the active elements in the living body into inorganic substances. The existence of the living body depends upon its breathing and the thermic processes going on within it, and it begins to disintegrate when these cease after its death. So the Greeks presumed that the existence of inorganic substance, with its various qualities and characteristics, depended upon a dynamic principle by which it was permeated. It was therefore natural to suppose that the active elements, or pneuma, were the basis of the existence of the inorganic world, just as they appeared to be responsible for the coherence and existence of the regular structure of any and every living thing. In man, this principle of life is the soul, often identified with breath. Thus, in the first instance, the all-permeating pneuma makes the whole living world a single unit, the difference between soul and organic life being merely the result of the variations in the composition of the pneuma. "According to (the Stoics) the soul is a pneuma, as is also organic nature. Only the pneuma is moister and colder in nature, and drier and warmer in the soul Thus it is a kind of primordial matter very similar to the soul, and the form of the primordial matter comes about through the mixture of the airy and fiery substances in suitable proportions" [192]. In taking the final step from the organic to the physical world the Stoics assumed that the pneuma fills the whole universe, both the space between bodies and the bodies themselves; it permeates all substances and makes them coherent, just as it expands in the space between the bodies. "Chrysippus' theory of mixture is as follows: he supposes that the whole of nature is united by the pneuma which permeates it and by which the world is kept together and is made coherent and interconnected" [193]. (We shall later speak in detail about the nature of this mixture according to the Stoic view.) This power of making coherent is one of the basic qualities of the pneuma, one of the signs of its activity which point to its derivation from the active elements.

"Those who have most expounded the concept of the binding force, like the Stoics, distinguish this binding force from what is bound together by it. The substance of the pneuma is the binding agent, while the material substance is what is bound together by it. They therefore say that air and fire bind together, whereas earth and water are bound together" [194]. This power of holding together the parts of the body so that they do not disintegrate is explicitly denied to the passive elements: "Pneuma and fire bind together themselves and everything else, whereas water and earth need something else to bind them together" [195].

How do the active elements, and the pneuma which took their place, come to have this cohesive force? It derives from the tensional qualities in them. At first this tension meant no more than the manifestations of the pressure of compressed air or the expansive force of steam from boiling water. However, the development of this scientific concept, by a compulsive inner logic of its own, turned this tension into a quantity characteristic of the inner cohesion of the substances and their degree of stability. "The Stoics say that earth and water have no binding force of their own, nor can they bind other substances together. They maintain their unity by partaking of the power of pneuma and fire. Air and fire on the other hand, through their inner tension and through being mixed with the other two, provide these with tension, permanence and substantiality" [196]. This tension (tonos), or the "tension of pneuma" (pneumatikos tonos) as it is explicitly called in many instances, is the most significant distinguishing quality of the pneuma, by force of which it becomes an entity not altogether unlike the concept of a physical field in contemporary science. In virtue of its dynamic character, tension gives a certain definite shape to all physical phenomena. Plutarch quotes an instructive passage from Chrysippus' numerous writings: "Passive and motionless matter is the substratum of the qualities, while these qualities are pneumata and aerial tensions inherent in the parts of matter and determining their form" [197]. The sentence which precedes this passage is still more explicit: "The structure of matter is simply air, for bodies are bound together by air. Likewise all that is bound together in a material structure derives its quality from the binding air which in iron is called hardness, in stone thickness, and in silver whiteness." This second function of the pneuma—

giving all the various substances the stamp of their specific qualities—is of particular interest since it is the continuum theory's counterpart of the functions performed by shape and arrangement in the atomic theory. We have seen that individual atoms were devoid of all the "secondary" qualities inherent in the bodies which they combine to form. These qualities were simply the product of spatial arrangement—that is to say, they were determined by combinations of certain shapes in a certain spatial order. Now, clearly, a consistent continuum theory requires another principle of differentiation: for here shapeless, undifferentiated matter can be infinitely divided and contains no definite unit on which a system of arrangements could be erected. The passages quoted above plainly show that the Stoics replaced the principle of *arrangement* by the principle of *synthesis* and used the latter in a double sense. First, the mingling of pneuma with matter transforms it from shapeless matter into a substance with definite physical attributes. Further, every one of these attributes depends for its quality and degree on the extent to which fire and air are mixed in the pneuma. All the special physical attributes of the substance are determined by the proportion of these two components in the pneuma. This is the interpretation to be given to the sentence already quoted: ". . . the form of the primordial matter comes about through the mixture of the airy and fiery substances in suitable proportions" [192]. Corresponding to a continuous and infinite gradation of physical attributes the Stoics defined a continuous and infinite gradation of syntheses of pneuma: by giving up the simplicity of the pneuma they solved the problem of differentiation within the continuum theory of matter. Those of the later commentators who opposed the Stoics naturally seized on this point and made great play with the fact that the principle of synthesis contraverts the concept of simple substance. "If the pneuma is formed by fire and air and permeates all substances, mingling with all of them, and if the existence of every one of these depends on it—how can a body still be something simple?" [198]. To-day we know that every self-consistent scientific system can only be achieved by compromises of this kind.

The ancient commentators on the Stoics had already remarked on the similarity between the pneuma and the aether. They meant no more than that both were rare and tenuous, for the place of the aether, Aristotle's "fifth substance", was in the skies and it was the substance of which the stars were made. It was only in a later period that the aether gradually came to occupy all the vast expanses of the cosmos. Then its main function was to fill the whole universe in conformity with the old dictum, "Nature abhors a vacuum." At the beginning of the modern scientific era the aether theory underwent transformations which made it very similar to the pneuma theory of the Stoics. The cohesive function, if no other, was ascribed to the aether in the most unmistakable terms by Newton, who at various periods of his life conceived of something like a "Unified Field Theory" to provide a single explanation of the phenomena of light and gravitation. At the end of the third part of his *Principia Mathematica* (1687), Newton adds a few remarks concerning "a certain most subtle spirit which pervades and lies hid in all gross bodies; by the force and action of which spirit the particles of bodies attract one another at near distances, and cohere, if contiguous". But cohesion was not the only quality attributed to the aether by Newton, as we see from what follows: "And electric bodies operate to greater distances as well repelling as attracting the neighbouring corpuscles; and light is emitted, reflected, refracted, inflected and heats bodies; and all sensation is excited, and the members of animal bodies move at the command of the will, namely, by the vibrations of this spirit, mutually propagated along the solid filaments of the nerves, from the outward organs of sense to the brain, and from the brain into the muscles." There is a close resemblance here between Newton's aether and the pneuma, a resemblance which extends to their functioning in the sphere of psychic phenomena. Newton does not explicitly state that the aether gives physical qualities to matter. But, in attempting to use the aether as an explanation for the force of gravity, he was compelled to follow the Stoics in giving up the simplicity of this primordial matter. He postulated different degrees of rarity and density for the aether, according to whether it is within matter or outside it. These variations were meant to explain the universal mutual attraction of substances (cf. his *Opticks* and his letter to Robert Boyle, 1678).

The atomists held that the only form of motion was the locomotion of particles, and the only way in which physical activity could develop was by the impact of particles upon each other. The Stoics were the first, thanks to their thorough-going conception of the continuum, to give careful thought to the nature of the propagation of a phenomenon in a continuous environment. Here we must note an appreciable advance upon the continuists who preceded them, especially Aristotle. In various places, such as his book *De Anima*, Aristotle discusses the phenomenon of sound and its propagation, but he nowhere mentions how it propagates. The Stoics were the first who explicitly laid stress on circular propagation in two, and spherical propagation in three, dimensions. They were also the first to use the classic analogy of water waves: "The Stoics say that the air is not composed of particles, but it is a continuum which contains no empty spaces. If it is struck by an impulse it rises in circular waves proceeding in straight sequence to infinity, until all the surrounding air is stirred, just as a pool is stirred by a stone which strikes it. But whereas in this latter case the movement is circular, the air moves spherically" [199]. There is also one explicit reference to hearing in connection with wave propagation: "We hear because the air between the voice and the hearer is struck and expands in spherical waves which reach our ears, just like the waves in a pool which expand in circles when a stone is thrown into it" [200].

It was obvious to the Stoics that this wave expansion within a continuous environment was connected with the tensional quality of that environment. The Stoic astronomer Cleomedes (*c.* first century A.D.) says: "Without one binding tension and without the all-permeating pneuma we would not be able to see and hear. For the sense perceptions would be impeded by the intervening empty spaces" [242]. The whole Stoic theory of perception was built upon the concepts of tension and the qualities of a continuum in a constant state of tension. In the Stoic view sight is due to the light which leaves the centre of the beholder's soul through his eye and connects him with the object seen when it reaches it and touches it. The mechanics of this process are described in one fragment as follows: "According to Chrysippus . . . sight is due to the light between the observer and the object observed being stretched conically. The cone forms in the air, with its apex in the eye of the observer and its base in the object observed. In this way the signal is transmitted to the observer by means of the stressed air, just as (by feeling) with a stick" [201]. The sensory perception of the object observed travels through the eye of the observer to his soul. In this operation, too, the pneuma plays a part, since it continues to operate as an "elastic medium" in the human body even in the case of mental processes. What is so remarkable here is the Stoics' profound understanding of dynamic phenomena in an elastic medium and their power of precise formulation. Though the problem first presented itself in connection with a biological phenomenon, its general significance for inorganic matter as well was immediately grasped. The Stoics started from their theory about the two-fold motion of sense perception from the centre of man's soul to his eyes or ears and back again, via contact with the object outside him. This picture calls to mind a physical phenomenon which was undoubtedly known to the Stoics, namely the expansion of a wave in a *confined* environment, such as a pool or tub. Here the waves expand in concentric circles around the stone thrown into the water, until they are thrown back from the sides of the tub or pool; thereupon, the returning waves interfere with those expanding outwards from the centre. This interference results in what is called in scientific language a "standing wave" or a "standing vibration". This kind of vibration is typical of any confined body, such as a vibrating musical string or bell. We may assume that for the Stoics it came to symbolize the co-existence of motion and rest in the same single system. For this vibration they coined the special term "toniké kinesis" (tensional motion). In the toniké kinesis of the pneuma they discovered a concept which successfully answers to the requirements of the continuum theory in explaining the qualitative differentiation of all the various organic and inorganic substances: every "self-vibration" imposes a quality of its own. This was the perfect counterpart of the Epicurean association of atoms, where differentiation resulted from the movements of groups of atoms set in vibration by repeated collisions. "There are those, like the Stoics, who say that there is a toniké kinesis in substances which moves simultaneously inwards and outwards. The outward movement gives rise to quantities and qualities, while the inward movement produces unity and substance" [202]. This concept puzzled many of the later commentators, for example Alexander of Aphrodisias: ". . . . What is the meaning of that movement which exists simultaneously in two opposite directions, outwards and inwards, and gives all objects their coherence which they call pneuma? How does this form of movement originate?" [198]. On the other hand, writers like Philo used this analogy to provide a single explanation for the structural differentiation alike of organic and inorganic substances. "He endowed some substances with physical structure, some with the power of growth, some with soul and some with an intelligent soul. In stones and trees which are divorced from the organic creation, he formed the physical nexus as a very strong bond. This is the pneuma that returns upon itself. It begins in the centre of the substance and stretches outwards to its edges . . . and returns back again to the place from which it started" [203]. This definition of the physical structure of a substance by the vibrations of the pneuma calls to mind a similar line of thought in the modern continuum theory. It is further proof of a fundamental fact that we have already met with in the atomic theory: similar basic scientific conceptions necessarily forge for themselves the same moulds of thought and means of expression, regardless of the technical resources of the age. After all that has been said, we shall not be surprised that the toniké kinesis also served as a figure for describing the propagation of a *state* in the broadest sense of the term. Philo went even further and tried to use it to explain the expansion of a non-physical occurrence. Discussing the movement of logos in the sense of a spiritual cause (the word of the Creator, for instance), he says: "It does not move by change of place, i.e. by leaving one place and occupying another, but by means of tensional motion" [204].

A scientist of the stature of Galen recognized the great value of the idea of tensile motion and used it in his essay on muscular movements. In his opinion, the function of the muscles is to be explained by their movement. In that case, it is difficult to account for the muscular movement in an arm extended but at rest. "Since neither the whole limb of which (the muscles) are part, nor they themselves appear to move, we do not dare to concede that they do move. What is the solution to this problem? Possibly we shall find it in the theory of those movements which are called 'tensional'" [205]. In other words, it is the standing vibrations of the muscles that explain away the difficulty of movement in repose. Galen adds a most interesting analogy between the dynamic balance of a substance and a standing wave. He distinguishes an object which is at rest because no force is acting upon it from an object which is at rest as the result of a balance of two opposing forces. He regards the second case as "doubtful rest". As an example of this, he takes a man swimming against the current of a river: "When his strength is equal to the force of the current he remains always in the same place, not like a man who does not move at all, but because he moves forwards with his own motion just as much as he is moved backwards by the external motion. . . . Sometimes a bird in the sky appears to remain in the same place. Are we to say that it is at rest as if it were hung up there? Or that it moves upwards just as much as its weight would pull it downwards? This latter explanation seems more correct. . . . So it is possible that in all these cases the object moves now upwards now downwards, as it is affected by the opposite motions in succession. But because of the great speed at which the changes occur and the extreme rapidity of the movements, it seems to remain stationary in the same place" [205]. The last sentence makes it clear why Galen appends this passage to the analogy of the tensile motion of the muscles. He is all the time conscious of the object as a field of perpetual conflict between equal and opposite forces. This dynamic equilibrium is for him analogous to the standing vibration that consists of two wave-systems moving in opposite directions.

The pneuma and its tensile force which fill the whole of the cosmos are also the agents which create the unity and inter-connection of its various parts. The structure of the entire cosmos is orderly and harmonious. This order and harmony depend upon the existence of the pneuma. Hence the tension of the pneuma becomes a cosmic force. This extension of the function of pneuma to the wider conception of the general connection between the parts of the cosmos was completed in Poseidonius' doctrine. The central feature of this is the concept of sympathy, the term used for the interaction of all the substances in the cosmos. "In the cosmos there is no void, as can be seen from phenomena. For if the substance of all things were not held together everywhere, the cosmos could not have a natural orderly existence and there would be no mutual sympathy between its various parts" [242]. Obviously, there is a very great difference between Aristotle's denial of the void and the Stoics'. Their conception of the cosmos as a single unit, a homogenous organism, rules out the existence of a void by giving a new revolutionary significance to the continuum concept. Instead of being represented as a mathematical and topological "juxtaposition", the continuum now appears as an "interrelation"—a physical field of activities and influences passing from place to place and from substance to substance and transforming the whole mass of entities into a structure which acts and is acted upon through the harmonious interpenetration of its parts.

The concept of sympathy and the idea of "the existence of union and tension between the celestial and terrestrial things" [206] are most strikingly demonstrated in Poseidonius' discovery of the connection between the moon and the oceanic tides. Poseidonius was a native of Rhodes. But he was a great traveller, and in the course of his journeys reached the Atlantic Ocean. At Gadeira in Spain (the modern Cadiz) he carried out accurate observations of the dependence of tides on the phases of the moon. He also collected information on this subject from the local inhabitants. The geographer Strabo, who lived a few decades after Poseidonius, describes the latter's findings in the third part of his *Geography*: "He says that the ocean has a periodic motion like the stars. There is a daily, a monthly and a yearly period, all due to sympathy with the moon. When the moon rises the breadth of a zodiacal sign above the horizon, the sea begins to rise and encroaches noticeably upon the dry land, until the moon reaches the middle of the sky. Conversely, when the moon descends the sea slowly withdraws, until the moon is the breadth of a zodiacal sign from its setting. In this condition the sea remains until the moon reaches its setting point and in its continued motion travels the breadth of a zodiacal sign from the horizon under the earth. From this point onwards the sea again rises until the moon's nadir under the earth. After that there is an ebb once more, until the moon climbs up to the breadth of a zodiacal sign below the horizon. This condition continues until the moon rises the breadth of a zodiacal sign above the earth, when the tide again rises. This is what he calls the daily cycle" [244]. Poseidonius' description of the monthly cycle further shows that he had also discovered the influence of the sun on the tides: "The monthly cycle means that the high tide is strongest at the new moon. It declines until the first quarter, then again increases until the moon is at its full. It again declines until the third quarter, after which it increases once more until the new moon." A more detailed, but essentially similar, description is given by Pliny the Elder, about a hundred and fifty years after Poseidonius. Poseidonius' contemporary, the astronomer Seleucus of Babylon, also knew of the connection between the moon and the tides. His explanation, however, is purely mechanical: the moon's motion deflects the air between it and the earth and it is this deflection that causes the high tides. Poseidonius explained the phenomenon by sympathy, that is to say, by the tension of the pneuma which makes possible the remote interaction of the moon and the earth. It must be admitted that this theory was as adequate as any could be at a time when the nature and universality of the force of gravity were still unknown.

We remarked at the beginning of this chapter that the term "pneuma" has a long history. The significance acquired by it in its last, Stoic metamorphosis raises a problem which calls for some discussion here. The pneuma is a mixture of air and fire, or at least some kind of product of the air alone. There can therefore be no doubting its materiality. Now, one of the few fundamental laws that have retained their validity throughout all the wreck-strewn history of physical concepts is that "where there is one body, there cannot be another". Hence, even if we suppose that the pneuma is composed of the most rarefied material, as extremely rarefied as aether was said to be in the seventeenth century and later, it will still necessarily be subject to this fundamental law. How then did the continuists picture the permeation of substances by the pneuma? This permeation must obviously be complete, since in the last analysis it decides the nature, cohesion and all the physical qualities of the substances. Is it enough to speak of a simple mingling of the material with the pneuma? And what is meant by such a mingling? This brings us to a problem which exercised the minds of ancient Greek scientists not a little and caused them no small headache—the problem of mixture. How are we to classify what results from mixing two or more different components? Aristotle discusses this question in the tenth chapter of the first part of his book on *Generation and Corruption*. He has a much harder task than the atomists. They, with their molecular conception of matter, could regard the structure of or the mixing of two liquids as a kind of mosaic made from the combination of the atoms of the two components, like a heap of wheat and a heap of barley mixed together. This is the illustration used by Aristotle to show that a "true" mixture, by definition, cannot be achieved in this way. "They say . . . that grains of barley are mixed with grains of wheat, if each grain lies beside one of the other kind. But every substance can be divided. Therefore, since a substance mixed with another substance is by composition equal in all its parts, it follows that every individual part of each component should lie beside a part of the other component. But there is nothing that can be divided into its ultimate parts, and combination is not the same as mixture, but different from it. Hence, components that have been mixed in particle form cannot be called a mixture, for this is a combination and not a mixture, and each part of the result will not show the same proportion between the components as the whole substance. But we have already found that a mixture must be equal in all its parts, if the components really have been mixed together. Just as every part of water is water, so is it with a mixture" [165]. Thus, Aristotle found a further flaw in the atomic theory: it would destroy the concept of a mixture by turning a macroscopic mixture into a combination of microscopic entities. On the other hand, neither is the continuum theory followed by Aristotle any better able to provide the microscopic homogeneity required to preserve the concept of a mixture, as long as it is assumed that in it each of the components will maintain its identity unimpaired. Aristotle tries to surmount the difficulty by distinguishing "actual" from "potential" existence. "Since some things exist actually while others exist potentially, it is conceivable that the components of the mixture exist in this latter sense, without actually existing. Actually the mixture will be different from its components, whereas potentially each one of them will be just as it was before the mixing and both will exist indestructibly" [165]. To make this conclusion more plausible, Aristotle argues that each component acts on its opposite number, changing it into something like itself, and that mixture results from this mutual assimilation, i.e. from a truly homogeneous compromise between the two. Thus Aristotle attempted to construct some sort of continuum theory as the basis of a doctrine of chemical unification, a process essentially different from mechanical mixing. But this example of interaction of the components and their mutual assimilation led him seriously astray. If the quantity of one component is much greater than the other, Aristotle continues, then "the result is not a mixture, but the increase of the dominant component; for the other material is transformed into the dominant component. Hence a drop of wine does not mingle with ten thousand pitchers of water; instead, it loses its identity and merges with the whole volume of water." Aristotle does not advance any arguments in support of this peculiar theory. Nor does he indicate any dividing-line below which one component merges into the other and above which the mixture is preserved by their mutual assimilation. Is this line fixed by outward marks? Can it be measured by our senses? Are we to say that as long as the wine colours the water there is interaction, and as soon as no change of colour is observable the wine is lost in the water? It looks as though Aristotle forgot Anaxagoras' warning about the weakness of the senses and his famous illustration of how this weakness keeps us from ascertaining the truth: our inability to observe the slight change which occurs when black paint is added drop by drop to white.

However that may be, it is clear that finding a solution to the problem of mixture compatible with the continuum theory is no easy matter. The Stoics, too, grappled with this problem. They were driven to do so, as has been remarked, by the whole question of the pneuma and its tensional permeation of matter. The Stoics were more methodical than Aristotle in their treatment. First of all they distinguished three kinds of mingling: mechanical intermingling, as in the case of seeds; mixture, in which liquids unite with liquids, or liquids with solids, while preserving their qualities; and finally, fusion which unites materials by the destruction of their own qualities, as in the case of drugs. ". . . . The quality of the component liquids, such as wine, honey, water, vinegar, etc., is evident in the mixture. Indeed, the quality of the components in these mixtures is preserved, as is shown by the fact that they can frequently be separated out by artificial means. . ." [207]. The identity of the component materials is preserved even in extreme cases where fractions of one kind are mixed with large quantities of another. This the Stoics had discovered from experience: "This can be seen in the case of frankinsense. When burnt in a flame it becomes greatly rarefied, but for all that it preserves its quality. There are many substances which, when assisted by other substances, expand far more than they could themselves. Gold, for example, when mixed with drugs or medicines can be rarefied far more than is possible when it is beaten out. . . . Hence, according to (the Stoics), we should not be surprised that there are substances which assist each other by forming a total mixture with each other. In this way their qualities are preserved in this thorough interpenetration, even if the mass of one is so slight that by itself it could not preserve its own quality in such a rarefied condition. Thus, for example, a small measure of wine mixes with a large amount of water, being helped by the latter to spread throughout such a great volume" [208]. This is the diametrical opposite of Aristotle's view: a mixture cannot become a fusion, even if one of the components is negligible in quantity. Chrysippus explicitly alludes to Aristotle's example of the drop of wine in pitchers of water when he says: "There is nothing to prevent one drop of wine from mixing with the whole sea. . . . By means of the mixture the drop spreads throughout the whole cosmos." [209]. Observe with what consistency the view is worked out: a change in the proportion between the component quantities cannot bring about any change in the nature of the mixture. The quotation before the last contains the expression "total mixture", used by Chrysippus to denote the mechanism of mixture and fusion as opposed to the process of combination. This is certainly an infringement of the principle that one substance cannot occupy the place of another. But only by such an extreme conception could the Stoics explain what happens in all the cases which in modern terminology would be called mixture, diffusion, suspension, alloy and chemical compound. "Total mixture brings about the expansion of the smallest quantities in the largest quantities, up to the confines of the substance. Whatever place is occupied by one is occupied by both of them together" [210]. Total mixture means the simultaneous existence of two substances one within the other. Their qualities may be either preserved or cancelled. Juxtaposition, on the other hand, "is the combination of substances by their surfaces, as in the case of heaps of produce containing wheat, barley, lentils, and similar things or pebbles and sand on the shore" [207]. It was the concept of total mixture that made it possible for the Stoics to develop their idea of pneuma: pneuma as permeating the whole cosmos, including the substances in it, and determining the qualities of these substances by mingling with them, just like that interpenetration occurring when "substances . . . stirred into a mixture interpermeate each other, and such that not a particle among them does not contain its share of all the rest" [208].

The whole controversy about the nature of mixture shows that the problem of continuity is closely connected with that of infinite division. The mixing of particles of finite size, no matter how small, is an elementary matter: it can be treated as the problem of the contact between the particles. But the difficulties begin when we consider particles that can be divided *ad infinitum*. Or, rather, they commence even earlier, with the problem of the infinite division of a continuous quantity. How did the philosophers and scientists whose view of the world was dominated by the continuum concept approach the problem of the infinitely small? The questions under discussion here belong to the realm of mathematics. They were posed in a dramatic and provocative way about two hundred years before Chrysippus by Zeno of Elea, a pupil of Parmenides. Zeno's famous paradoxes go straight to the root of the matter. Plato was inclined to dismiss them as logical acrobatics. Aristotle, on the contrary, recognized their importance and dealt with them at length in the sixth part of his *Physics*, without however satisfactorily solving them. Here we shall examine three out of the four paradoxes.

"Zeno set four problems about motion which are hard to solve. The first concerns the non-existence of motion, since the moving body has to reach the half-way point before it reaches its goal. . . . The second, called 'Achilles', is that the fastest runner will never overtake the slowest, since the one behind must first reach the point from which the one in front started, so that the slow runner will always remain in front. . . . The third is . . . that the arrow in flight is at rest. This results from the assumption that time is composed of points of the present; without this premiss the conclusion will not follow" [68]. As Aristotle remarks, the only real difference between the first and second problems is that in the former the goal is fixed, whereas in the race it is constantly advancing. It will therefore be sufficient for us to examine the first and third problems, in so far as they throw light upon the Greek approach to the problem of a continuum. The first paradox can be represented diagrammatically by a section of a straight line equal in length to unity. In order to travel from its left end, i.e. zero, to its right end, i.e. a point at a distance of unity from the first, we have to pass an infinite number of points obtained by successively dividing into two first the whole section, then its right-hand half, then the right-hand half of this half, and so on. The distances of these points from zero are: $\frac{1}{2}, \frac{3}{4}, \frac{7}{8}, \frac{15}{16}, \frac{31}{32}, \frac{63}{64}\ldots$. This is an infinite series of numbers, all of which are smaller

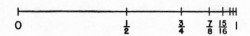

FIG. 8. One of Zeno's paradoxes: infinite dichotomy.

than one. That is to say that it is impossible to reach the goal—the right end of the line—in a finite number of steps. Hence, even before we begin to treat Zeno's paradox as a problem of motion to be solved purely by arithmetic or geometry, we are confronted by all those difficulties whose solution calls for a new form of mathematical discipline. This discipline is the infinitesimal calculus. This was first established in the seventeenth century, though its fundamental concepts and methods did not receive a fully satisfactory scientific exposition until the second half of the nineteenth century. To-day, one of the terms most commonly used by every student of higher mathematics is the "limes", i.e. the limit converged upon by the members of an infinite series, such as that given above. No single member of that series will reach unity—the right end of the line. But the difference between unity and those members grows smaller and smaller. We can bring it as close to zero as we wish, simply by introducing into the series a sufficiently large finite number of members. Obviously, there will always remain an infinite number of points between any dividing-line and the goal, as the concept of a continuum requires. No one of these points of division is the last: they go on infinitely, crowding closer and closer together near the goal. This is the fact that makes possible that transition to the limit which reduces to zero the distance between the intermediate points and the right end of the line. The infinitesimal calculus has here introduced a dynamic element into mathematics, thanks to which the problem of the infinitely small can be adequately handled. The importance of Zeno's paradoxes lies in their showing the impossibility of resolving the dilemma by the static mathematical formulae evolved to deal with problems concerning finite numbers.

We shall subsequently see the influence of Zeno's profound ideas on Greek analytical thought. We shall see how, one hundred years later, Eudoxus of Cnidus—and a hundred years after him, Archimedes—used the principle of convergence in their geometrical calculations. We shall also observe how Chrysippus made this principle a main pillar of his continuum theory. But first we must return to Zeno's paradoxes. So far we have examined the first as if it were a purely arithmetical problem. But in fact it is presented as the kinetic problem of a body in motion. From this point of view, the situation is given in its most extreme form in the third paradox—that of the arrow which is at once in flight and at rest. It is in flight, for it moves from place to place in space; on the other hand it is at rest at every moment defined as a point of "now". Since there is once again an infinite number of such points between any two given points, the whole movement is apparently made up of states of rest.

Aristotle, who saw the problem clearly, could not find a satisfactory answer to it. He realized that it is not enough to regard time as made up of "nows", but that one has to take into account the "before" and "after", too (cf. his definition of time [147], already quoted earlier). On the other hand he knew that a limited portion of time is made up of an infinite number of points, just as a limited distance is. Hence the problem of the third paradox is essentially the same as the first: how is it possible for the arrow to have a motion that carries it over a finite distance in a period composed of infinite points of time? In his analysis of the first paradox (*Physica*, Chap. VI), Aristotle came very near to the solution without actually reaching it. "For this reason Zeno's paradox is wrong, when he supposes that nothing can pass things of infinite number, or touch them one by one in a finite time. For distance and time, and everything continuous, are called 'infinite' in two senses: either as capable of division, or as regards (the distance) between the extremes. It is not possible for anything to come into contact in finite time with things that are in extension. But this is possible when they are infinite in division.

Indeed, in this sense time itself is infinite" [149]. Aristotle here regards distance and time as similar: though both are of finite length they can both be divided infinitely. It was a correct intuition of his to couple them together in this way, since the problem of motion cannot be solved except by connecting time and distance together through a physical quantity well known to all of us from our schooldays—velocity. By definition, velocity is the ratio of the distance traversed by a body to the time taken. It will be seen that in this definition *two* points in space and *two* points in time are required, i.e. the two ends of the distance covered and the two ends of the relevant portion of time. Even this elementary concept was never accurately defined by Aristotle and the ancient Greek scientists: it is the creation of modern physics. To-day we are accustomed to speak of the velocity of a body at a given *point*. We say, for example, that a car passed a given line in the street at a speed of 40 m.p.h. How can we apply a kinetic concept requiring two points of space and time to a single point? The answer to this question is the answer to Zeno's paradoxes of motion. It was given by Newton, two thousand one hundred years after Zeno. If we consider two points close to each other on the arrow's path, the corresponding points of time will also be close together. If we now use the "dynamic" concept of the transition to the limit and consider ever-decreasing distances, i.e. two points the distance between which tends to zero, then the distance between the corresponding points of time will also tend to zero; but the *ratio* between these two distances, i.e. the velocity at the coalescence of the two points, will tend to a given finite quantity. In this way it is possible to speak of a definite velocity at a single point which is, as it were, one of the body's points of rest.

We shall leave to the last chapter the question why the Greeks never achieved even an elementary definition of so fundamental a concept as velocity. But from what has been said we can understand how much greater are the difficulties involved in coping with the continuity of time as compared with that of space. Time is not only indispensable for the description of motion but indeed of every form of physical action as an independent variable on which all the physical quantities depend. Unlike the transition to a limit for purely spatial quantities, which was fully grasped by some Greek mathematicians and above all by the Stoic philosophers, the convergence of points of time involves a clear notion of functional dependence. Here even the Stoics failed to overcome the analytical difficulties. However, the relevant texts prove that they had a deeper insight into the implications of the paradox of the arrow than Aristotle. This is first shown by their definition of time, which was stated in almost similar terms by Zeno the Stoic and by Chrysippus. The latter said: "Time is the interval of movement with reference to which the measure of speed and slowness is always reckoned" [211].

Here the connection with velocity is clearly indicated, whereas the term "interval" might be a hint that the necessity for defining two points of time was already realized by these philosophers. They were unable, however, to make the "dynamic" transition to smaller intervals coalescing in one point, and therefore cut the Gordian knot by defining the "now" as follows: "The Stoics denied the existence of a shortest portion of time, since 'now' is an indivisible quantity, and what is regarded as existing in the present . . . is distributed with one part over the past and the other over the future" [213]. Plutarch, moreover, quotes Chrysippus, who in his book *On Parts* stated the axiom that "part of the present is in the past and part in the future" [212]. This formulation, with its definition of the present as the centre of a very small, but still finite, portion of time, is clearly an attempt to comprehend the elements of time as finite "quanta" and not as extensionless points. The present thus becomes, so to speak, an "atom of time", or, to use the language of calculus, a differential of time. Plutarch, in his criticism of Chrysippus' solution, indicates its repercussions on the description of physical entities depending on time: "As regards actions and movements this leads to a complete confusion of clarity. For necessarily, if 'now' is distributed over past and future, what moves now will be something which partly has moved and partly is going to move . . . and likewise what acts will be a thing that partly has acted and partly is going to act" [214].

So great was the desire of the Stoics to give a clear answer to the paradox of the arrow, that these bitter opponents of the atomic hypothesis and ardent champions of continuum and no compromise had to have recourse to an "atomic" solution. The reasons given by Chrysippus are reported by a source already quoted above: "He says that it is most evident that there is no time existing in the present. For the dissection of all continuous things goes on infinitely and, by the same process of division, time is infinitely dissectible too; therefore no time exists in the present in a precise way, but is defined only broadly" [211]. To-day we can appreciate all the more the boldness of Chrysippus' solution, as we have witnessed modern attempts at circumventing "infinity catastrophes" in physics by a similar recourse to atomic concepts of length. Suggestions such as to solve the problem of the infinite proper energy of the electron by introducing a "minimum length" are—seen in the right perspective of the history of ideas—akin to that of Chrysippus.

It is perhaps also worth our while to bear in mind that the obstacles to the analytical description of time on the analogy of length are great, bound up as they are with our own inner consciousness of time: our fundamental biological sensation of the onward flow of time stands in the way of its comprehension in the abstract and its transformation into a geometrical dimension. This decisive step, which was the prerequisite of the evolution of modern physics, had to wait for Galileo.

The Stoic approach to the problem of the spatial continuum was entirely different, as we shall see presently when we return to the concept of "limes", the corner-stone of the higher calculus. Eudoxus in the fourth century and Archimedes in the third century B.C. made use of the principle of convergence in their geometrical proofs. Archimedes' example is famous: he based his calculation of π, i.e. the ratio of the circumference of a circle to its diameter, on the fact that the circumference is larger than the perimeter of every inscribed polygon and smaller than that of every circumscribed polygon. If we consider the sequence of regular inscribed polygons whose sides constantly increase in number, we find that their perimeters form a series of lengths constantly increasing but always less than the limit fixed by the circumference of the circle. Conversely, with the regular circumscribed polygons, as the number of their sides increases, their perimeters decrease, but always remain above the limit fixed by the circumference of the circle. In this way, the circumference of the circle can be intercepted between two series of lengths converging from opposite sides. Eudoxus used the same method to prove that the areas of two circles are related to each other as the squares of their diameters. We are also told by Archimedes that using the principle of convergence, Eudoxus successfully proved that the volume of a cone (or pyramid) is one-third of the volume of a cylinder (or prism) of the same base and height. Archimedes says that it was Democritus who discovered this theorem, but was unable to prove it. From Plutarch we learn how Democritus, apparently in the course of his search for the exact proof, came upon the problem which he formulates as a paradox: "Now see how Chrysippus answered the very real difficulty raised by Democritus in one of the problems of natural science. If a cone is cut by sections parallel to its base, are we to say that the sections are equal or unequal? If we suppose that they are unequal, they will make the surface of the cone rough and indented by a series of steps. If the surfaces are equal, then the sections will be equal and the cone will become a cylinder, being composed of equal, instead of unequal, circles. This is a paradox" [215]. Here we have yet another problem that cannot be solved by static concepts. Democritus saw no way of building the cone from circular

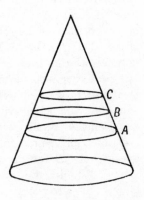

FIG. 9. Democritus' paradox of adjacent parallel sections of a cone.

segments each of which would be slightly different in area from the one above or below it. The father of the atomic theory, with his belief in finite units, understood this "slightly" as a finite difference resulting in a series of steps. If, on the other hand, the difference were infinitely small, it would be non-existent and the result would be a cylinder. The passage from Plutarch quoted above contains two answers by Chrysippus to the problem posed by Democritus, both of which are most interesting as an attempt to create a terminology for the infinitely small. Chrysippus says: "Sometimes one thing is larger than another without protruding" [216]. This at once calls to mind the differential concept: when a differential is added to a quantity, the quantity increases, but the increase is infinitely small, it "does not protrude". In other words, the tendency to zero of the distance between the two sections brings with it a tendency to zero of the differences in their areas, with the result that we get a perfectly smooth cone.

The expression "larger without protruding" adequately represents the modern mathematical term "greater or equal". In this connection Chrysippus' second formula is of interest. His opponents no doubt posed him the following problem. If we take three contiguous sections of the cone and mark them from bottom to top A, B, C, then the segment bounded by the sections A and B will be a body larger than that bounded by B and C. Now, if sections A and C are brought near to section B, A will not be noticeably larger than B and B will not be noticeably larger than C. Hence in transition to the limit A will not be noticeably larger than C and we shall once more get equal bodies, i.e. a cylinder instead of a cone. Chrysippus's answer, as quoted by Plutarch, is as follows: "The areas will be both equal and unequal, but the bodies will not be equal, since their areas are both equal and unequal" [215]. The expression "areas at once equal and unequal" points to the infinite series of sections A in their approach to B (or of sections C in their approach to B from above). When bodies are bounded by areas like these which are larger than a given area or equal to it in the convergence from both of its sides, the bodies differ in volume. Plutarch fails to understand the dynamic notion implicit in the strange expression "equal and unequal". He therefore finds here an infringement of the fundamental concepts of logic and tries to represent this whole approach as fantastic. In actual fact, the Stoics here hit upon the definition of infinitesimal quantities, thus theoretically opening the way to the development of the higher calculus in antiquity. That such a development did not follow was solely due to the inability of the Greeks in this sphere to translate ordinary speech into mathematical symbols.

There is a connection between Eudoxus' method of proving the theorem that the volume of a cone is one-third that of a cylinder and the whole dispute between Chrysippus and Democritus. Eudoxus divided the cone into many segments by means of parallel sections, and then inscribed each section between series of circumscribed and inscribed cylinders. These cylinders formed a series of steps enveloping the smooth cone. The required demonstration is by the summation of the volumes contained by the series of steps and by convergence. It is obvious that such proofs and the use of the principle of convergence by mathematicians like Eudoxus and Archimedes were known only to a select company of "experts". For this reason, it is particularly significant that a philosophical school of the standing of the Stoics should have included this principle in its doctrine and, by formulating it in general terms, should have given these concepts a far wider publicity. The method of inscribing a quantity between two convergent series, for example, was formulated in the most general terms by the Stoics and applied to any "body" whatsoever. Remembering that in Stoic terminology the word "body" is of very wide application, embracing geometrical quantities, time and space, substances and even qualities, we can regard the following definition as a very general expression of the dynamic

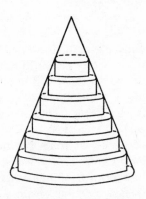

Fig. 10. Eudoxus' approximation of the volume of a cone by summation of the volumes of inscribed cylinders.

character of the continuum: "There is no extreme body in nature, neither first nor last, with which the size of a body comes to an end. But every given body contains something beyond itself and the substratum is inserted infinitely and without end" [217].

It is very possible that this definition is also meant to refute the arguments employed in the futile controversy about the position occupied by a given body in space. Is the surface of the body part of its "place"? Or does it merely delimit "place" without being part of it? The Stoic definition exposed the futility of such hair-splitting by providing the terminology necessary for representing every quantity in the continuous cosmos as a limit converged upon by its contiguous quantities.

By the third century B.C. the Greek understanding of infinity had gone far deeper than the mere invention of a device for solving mathematical problems or the coining of a philosophical aphorism. This is evident from the fact that simultaneously with the deeper comprehension of the infinitely small, advances were made in the realm of the infinitely great. With regard to the infinitesimally small, the central concept, from which radiate all

155

other concepts and functions, is convergence. In the case of the infinitely large this central position is occupied by the concept of the set. The characteristic difference between a finite and an infinite (large) number finds expression in the difference between a finite and an infinite set. "Set" is a collective term for units of a given kind. If the set is finite, part of it is always less than the whole, as can be shown by pairing the units of the part with the units of the whole. In this case there results an excess of units in the whole set which have no partner in the partial set. But if the group is infinite, a different situation arises. This was first proved in detail by Galileo, who devotes a number of pages to this problem in his book *Discourses and Mathematical Proofs,* published in 1638. But it was not until the nineteenth century, with Cantor's theory of sets, that the problem became an important branch of mathematics. Two simple examples will make clear the characteristic quality of an infinite set. Let us take an infinite numerable set such as that of the natural numbers. A partial set of this (also infinite) is formed by all the square numbers. Now it is easy to show that the part is equal to the whole, by placing a natural number side by side with its square. Every natural number without exception can be paired with a square number. Hence the capacities of the whole and of the partial set are equal. For the second example we shall take an innumerable set (i.e. one which cannot be counted like the natural numbers)—the set of the points included in a section of a straight line. Let us take a part of this section and superimpose it on a line parallel to the whole section. We then join together the left ends of the two sections and produce this line to meet the line joining together their right ends. Now every line drawn from this apex and cutting the two sections pairs off a point of the whole section with a point of the partial section. Hence the number of points in the part is equal to the number in the whole. These two superficial illustrations are sufficient to bring home to us the signi-

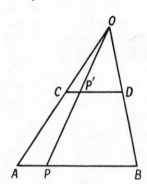

FIG. 11. The number of points on *AB* equals that on *CD*: each point *P* on *AB* can be paired with a point *P'* on *CD*.

ficance of the Stoics' achievement in being the first to define the infinite set. Their definition is once again quoted by Plutarch, who was so far from understanding its meaning and importance that he takes the Stoics to task for their nonsensical statements. "Is it not self-evident that a man is composed of more parts than his finger, and similarly that the parts of the cosmos are more numerous than those of a man? This was known and understood by everyone, until the Stoics came along. They stated the opposite thesis—that the parts of a man are not more numerous than those of his finger, nor the parts of the cosmos than those of a man. For, so they say, bodies can be divided infinitely and there is no large and small with infinite quantities, nor is one of them greater than another: the remaining parts never cease splitting up and providing a quantity out of themselves" [218]. These words plainly show that the Stoics were aware of the characteristic difference between a finite and an infinite set. This awareness, together with the passages quoted on the subject of convergence, prove that the Stoics succeeded to an astonishing extent in really coming to grips with the problems of a continuum and the questions of infinity inseparable from it. The explanation of this success is to be found in their dynamic conception of the continuum, as indeed of the whole cosmos. For them the cosmos was the most majestic embodiment of all those categories of continuity which maintain the stability of the whole creation through the eternal dynamic of its parts.

VII

THE INTERDEPENDENCE OF THINGS

"For the cause was from the Lord."
1 KI. 12.15

WE have seen how the two great rival schools of the Hellenistic era served as the mouthpieces of two opposed scientific views. While Epicurus and his disciples expounded the atomic theory, the Stoics, especially Chrysippus and Poseidonius, taught the continuum theory. As the teachings of these philosophical systems included not a little about many other fields of science, this led to a process of popularization of natural sciences within wide sections of the population. The writings of the "pure" scientists, and even the greatest of them like Archimedes, Eratosthenes and Hipparchus, presumably were confined to a small circle of mathematicians and astronomers. One has to bear in mind that the autonomous scientific research of those times remained fragmentary without ever achieving the continuity and tradition attained by modern science from the seventeenth century onwards. Thus the main channels for dissemination of science and scientific concepts were the philosophical schools which so deeply influenced the culture of the Ancient World.

As a result of this marriage of science and philosophy, scientific concepts were to some extent moulded by philosophical principles. It is, therefore, quite understandable that the confusion of doctrines and concepts and the tendency to admit into science principles and objectives from other spheres of thought often led

158

to consequences harmful to the development of scientific method and the clarification of fundamentals.

A proof of what has been said can easily be supplied from the attitude of Epicurean doctrine to the category of cause and effect in nature. The whole subject of the concepts used in Ancient Greece concerning determinism and causality needs to be treated with special care; first, for the general reason that we should avoid any uncritical projection of our own concepts into the past, and secondly because of the changes which occurred during that era in the meaning and usage of the terms in which that causality is expressed. The law of causality has acquired a specific meaning in modern times through the development of mathematical physics since Newton; and, despite the modifications to which it has been subjected by the Quantum Theory, that meaning remains essentially unchanged. There is, however, a more general meaning to the law of causality, a kind of supposition which is prior to any specific application of it. In its simple form this states that there is conformity with law throughout nature: nothing is arbitrary, there is a necessity for everything, as we see in the regular recurrence of all phenomena. Without this necessity, no accumulation of experience would be possible. The specific meaning of the law developed out of its general meaning as the most important of its uses. Its success in the realm of theoretical physics provides the fullest confirmation of the general law.

The conception of a general conformity with law existing in nature is contained in Greek philosophy from the beginning. The attempts of Thales and his followers to define primordial matter and to discover the fundamental laws underlying phenomena are simply an expression of this conviction. We are told about several of the pre-Socratic philosophers that they specifically expressed their belief in determinism. The extant fragment of Leucippus quoted in a previous chapter, plainly states: "Nothing occurs by chance, but there is a reason and a necessity for everything" [70]. This sentence, linking together reason and necessity as opposed to chance, displays great precision of thought. The conception of chance is connected in our minds with lack of logic and reason, because of its departure from the pattern of regular phenomena—or, more precisely, of phenomena which are permanently linked with other phenomena. It is this permanent connection, this continual repetition of the same sequence of occurrences, that gives to the "non-accidental" phenomena the character of necessity and the stamp of causality. This causality appears in its most perfect and purest form in the eternal cycle of stellar movements, though Leucippus' observation applies, no doubt, to all natural phenomena, to the entire cosmos, the permanence of which is evinced on a large scale by the recurrence of the same events, processes and their combinations. Even though the Ancient Greeks had little knowledge of experimentation, nothing more than the observation of nature was needed to lead them to Leucippus' conclusion. The opinion of Democritus, who followed Leucippus in this matter, is preserved by two sources. The first tells us: "Democritus of Abdera assumed that the universe is infinite, since it was not created by anyone; he likewise calls it unchangeable and specifically describes its qualities. There is no beginning to the causes of what is happening now, and all that was, is and will be has been contained in necessity from all eternity" [109]. The absence of change, which to Democritus appears characteristic of the cosmos, is simply the web of continually recurring things and the sequence of combinations repeated identically, or with modifications in which, too, permanence can be found. Hence his conclusion about the eternity of causes and their necessity, which gives rise to confident belief in a fixed order of things in the future.

The second source for Democritus' opinion on this question is Aristotle, who mentions it in order to refute it: "If we make the general assumption that we have found a satisfactory principle in the fact that something is 'always' so, or will 'always' occur in such a way, we shall be mistaken. Democritus, for example, builds his theory of natural causes on the fact that things happened in the past just as they happen now. But he does not find it necessary to search for a principle which will explain this 'always'" [110]. It is impossible for the modern physicist to find any justification for Aristotle's attack: he will most certainly approve of Democritus' position, almost prophetically based on an analogy unknown in his time—the analogy of the machine. The "necessity" in phenomena, their causality, can be expressed in terms of a mechanism, or part of a mechanism, the essence of which is that it always functions in the same way. Even modern developments in physics, though weakening the mechanical analogy, have not changed anything in this respect; the mechanism has been replaced by the mathematical equation and mathematical rules of operation. These, too, give expression to the same Democritean principle that something will "always" work in such a way; in other words, they still perform the function of the mechanism from which we observe how something "happened, happens and will happen" by reference to its special structure and the interdependence of its parts and on the basis of assumptions about the way in which the parts work. It is true that neither Democritus nor his successors in the classical world managed to translate this idea into mathematical terms, or to provide it with specific illustrations by means of our modern system of induction and experimentation. Still, it is not for this reason that Aristotle attacks him, but from a point of view which we cannot accept to-day: Aristotle rejects the whole system of explaining the phenomena of nature, physical and biological alike, on the analogy of the machine and the automatic mechanism. The analogy which he uses is the creative artist, whose creative work is directed to a definite end—the giving form to formless matter. Even if Aristotle had accepted the machine as part of his analogy, he would not have stopped at its description, but would have raised as fundamental the question of who invented the machine and for what purpose. Democritus—like the modern physicist—did not include this question in his picture of the universe but contented himself with the description of the universe as a machine in order to provide it with causality. In this, and in his refusal to be drawn into "metaphysical" questions, we can see his intellectual acumen and his deep understanding of physics. He and his teacher, Leucippus, may be considered the first to give scientific formulation to the causal law in nature.

The fact that the founders of the atomic theory, based as it is on concepts of pure mechanics, were also the first to formulate the idea of mechanical causality does not require any further analysis or clarification. But in the teaching of Epicurus we come to a parting of the ways, and, side by side with the progress in the atomic theory mentioned in Chapter V, we have to note a serious setback as regards the connection between cause and effect. This retrogression is twofold: Epicurus gives up the idea of a total prevalence of "necessity" in the cosmos, and at the same time he displays glaring inconsistency in the application of the principle of causal uniformity to a certain category of natural phenomena. The reasons for these two serious deviations from the clear path of scientific reasoning are to be found in the domination of scientific teaching at that time by philosophical thought, as has been indicated above. In the last resort, Epicurus did not take science seriously, and wherever, in his opinion, his conclusions were likely to endanger the spiritual quietude of man and to disturb his happiness by arousing fear or superstition, he prefers man to science. Belief in man's freedom of will was one of the fundamentals of Epicurus' teaching. Since this is not compatible with the absolute rule of necessity in the universe, with the conception of complete determinism to which, at that time, the term "fate" (fatum, heimarmené) was usually applied, Epicurus found it necessary to endow atoms with free will. As soon as this is introduced at any suitable point, the extreme mechanistic nature of the atomic theory, which assumes that the soul too is composed of atoms, will somehow ensure its connection with man's freedom of will. Epicurus was able to get round the problem by means of his own assumptions about atomic movements. In the beginning, before the molecules and all the other combinations of minute bodies had been formed by the collision and coalescence of atoms, the latter, according to Epicurus, had a single, uniform motion. This uniformity of velocity is understood by him just as the modern physicist understands it: constant size and direction. Thus in the beginning there were atoms and empty space, with the atoms all moving at a constant velocity, in a straight line in the same direction. What was this direction? This question exposes the weakness of the Epicurean cosmogony. His atoms, which possessed weight, "fell" in a certain direction, just as any heavy body that we know of falls, even though in that epoch before the creation the terms "up" and "down" seem to be meaningless. But even in this his weakest point, we find another astonishing flash of intuition: against Aristotle Epicurus categorically asserts that in the void the velocity of all the falling bodies is the same, regardless of their weight: " Moreover, the atoms must move with equal speed, when they are borne onwards through the void, nothing colliding with them. For neither will the heavy move more quickly than the small and light, when, that is, nothing meets them: nor again the small more quickly than the great, having their whole course uniform, when nothing collides with them either" [181]. This is, in principle, the conclusion which Galileo reached in his laws of falling bodies. Only thereby Epicurus raised a serious problem for his cosmogony: how can those collisions, without which the universe would never have been created nor any macroscopic bodies have been formed, come about? .

At this point "freedom of will" appears as a *deus ex machina*. This is how Lucretius describes the situation in his poem: "When the atoms are travelling straight down through empty space by their own weight, at quite indeterminate times and places they swerve ever so little from their course, just so much that you can call it a change of direction. If it were not for this swerve, everything would fall downwards like rain-drops through the abyss of space. No collision would take place and no impact of atom on atom would be created. Thus nature would never have created anything. If anyone supposes that heavier atoms on a straight course through empty space could outstrip lighter ones and fall on them from above, thus causing impacts that might give rise to generative motions, he is going far astray from the path of truth. The reason why objects falling through water or thin air vary in speed according to their weight is simply that the matter composing water or air cannot obstruct all objects equally, but is forced to give way more speedily to heavier ones. But empty space can offer no resistance to any object in any quarter at any time, so as not to yield free passage as its own nature demands. Therefore, through undisturbed vacuum all bodies must travel at equal speed though impelled by unequal weights. The heavier will never be able to fall on the lighter from above or generate of themselves impacts leading to that variety of motions out of which nature can produce things. We are thus forced back to the conclusion that the atoms swerve a little—but only a very little, or we shall be caught imagining slantwise movements, and the facts prove us wrong. For we see plainly and palpably that weights, when they come tumbling down, have no power of their own to move aslant, so far as meets the eye. But who can possibly perceive that they do not diverge in the very least from a vertical course?" [248]. Here Lucretius gives up the Democritean necessity and revokes the general application of the law of causality as it had been established by Epicurus' predecessors. He does this primarily to enable himself to trace the continuation of the

163

cosmogony which would remain without any such sequel if everything were governed "by reason and necessity". But this element of indeterminism immediately serves as a proof of free will, as appears from the following verses of Lucretius: "Again, if all movement is always interconnected, the new arising from the old in a determinate order—if the atoms never swerve so as to originate some new movement that will snap the bonds of fate, the everlasting sequence of cause and effect—what is the source of the free will possessed by living things throughout the earth? What, I repeat, is the source of that will-power snatched from the fates, whereby we follow the path along which we are severally led by pleasure, swerving from our course at no set time or place but at the bidding of our own hearts? . . . Although many men are driven by an external force and often constrained involuntarily to advance or to rush headlong, yet there is within the human breast something that can fight against this force and resist it. . . . So also in the atoms you must recognize the same possibility: besides weight and impact there must be a third cause of movement, the source of this inborn power of ours, since we see that nothing can come out of nothing. For the weight of an atom prevents its movements from being completely determined by the impact of other atoms. But the fact that the mind itself has no internal necessity to determine its every act and compel it to suffer in helpless passivity—this is due to the slight swerve of the atoms at no determinate time or place" [248].

It should be noted that the whole problem of the irreconcilability of determinism and free will became really acute only in the post-Aristotelian period, and especially, as we shall see, in the Stoic philosophy, whereas in Democritus there is still no indication of this dilemma. This shows how deeply causality, as an all-embracing law, had penetrated into men's intellectual consciousness in the little more than a hundred years from the end of the pre-Socratic period to the time of Epicurus and the Stoic Zeno. We are reminded of the long controversy on the same topic, starting with Laplace and continuing throughout the nineteenth century, which was carried on by both philosophers and natural scientists. This controversy has been opened up again in this century with the development of quantum mechanics and the formulation of the principle of "uncertainty". It is perhaps worth emphasizing, at this point, that there is no suggestion of any analogy between this principle and Epicurus' idea of "uncaused deviation". The modern problem was thrown into relief for us as a result of the experimental method in atomic physics and by the analysis of the interaction between the observer and the observed object. In the essentially non-experimental physics of Ancient Greece such a problem could never arise. Epicurus tries to get round his difficulty by means of the simple and somewhat primitive solution of removing one specific link from the infinite chain of physical causes in order to maintain the principle of free will as part of the cosmogonic process. Characteristic of Epicurus' rather superficial approach is his tendency to obscure the difficulties. On the one hand he emphasizes the minuteness of the deviation which is barely discernible; on the other hand, while stressing man's consciousness of free will as a fundamental reality, he tends to ignore all evidence of man's dependence on causal factors. All this harmonization is aimed at preserving, by every means, the peace of man's soul and avoiding any disturbance of its balance. In his letter to Menoeceus, Epicurus writes: "For, indeed, it were better to follow the myths about the gods than to become a slave to the destiny of the natural philosophers: for the former suggests a hope of placating the gods by worship, whereas the latter involves the necessity which knows no placation" [186].

This "myth about the gods", which is preferred by Epicurus to the nightmare of fate, played havoc with the Epicurean understanding of nature in another respect. The physics of Plato and Aristotle introduced the antithesis of heaven and earth by interpreting the eternal recurrence of heavenly motions as a sign of their divine nature (Plato), or as the most sublime instance of nature's striving for perfection (Aristotle). The marvellous conformity to law of all the heavenly cycles, which could be demonstrated in geometrical models or mathematical calculations, became a symbol of the divine and a conclusive proof of the existence of a supreme intelligence consciously controlling the cosmos. All this conflicted with Epicurus' line of thought. In his desire to set man free from fear of the gods without going to the extreme of atheism, he assigned to the gods a kind of symbolic existence of perfect happiness completely divorced from cosmic events and human destiny. This was his version of the Platonic ideal of the condition which a man can attain, if he preserves the peace of his soul. This is the key to the following excerpt from Epicurus' letter to Herodotus: "Furthermore, the motions of the heavenly bodies and their turnings and eclipses and risings and settings, and kindred phenomena to these, must not be thought to be due to any being who controls and ordains or has ordained them and at the same time enjoys perfect bliss together with immortality.... Nor again must we believe that they, which are but fire agglomerated in a mass, possess blessedness, and voluntarily take upon themselves these movements. But we must preserve their full majestic significance in all expressions which we apply to such conceptions, in order that there may not arise out of them opinions contrary to this notion of majesty. Otherwise, this very contradiction will cause the greatest disturbance in men's souls. Therefore we must believe that it is due to the original inclusion of matter in such agglomerations during the birth-process of the world that this law of regular succession is also brought about. Furthermore, we must believe that to discover accurately the cause of the most essential facts is the function of the science of nature, and that blessedness for us in the knowledge of celestial phenomena lies in this and in the understanding of the nature of the existences seen in these celestial phenomena, and of all else that is akin to the exact knowledge requisite for our happiness" [182].

These words, especially the emphasis placed on the stars being "masses of fire" whose laws of motion were determined with the creation, give the impression that Epicurus wishes to continue the scientific tradition of the first natural philosophers and of Anaxagoras and Democritus, with a view to including the phenomena of heaven and earth within a single system. But his fear of religion led him astray; and in order to challenge the scientific pride of place held by celestial phenomena on account of their regularity and exactitude, he adopts an arbitrary differentiation of his own, thus reducing the value of his whole scientific doctrine. On the one hand, as regards what is invisible to the naked eye, in all that concerns the atoms, he makes assumptions and draws conclusions with complete confidence in the uniformity and unambiguity of the scientific explanation; but, in the case of the celestial phenomena, which though distant from us are still visible to the eye, he refuses to lay down any fixed laws and postulates the principle of alternative explanations, which rules out any possibility of coming to a comprehensive scientific conclusion. A few passages from his letter to Pythocles will provide the explanation: "First of all then we must not suppose that any other object is to be gained from the knowledge of the phenomena of the sky, whether they are dealt with in connection with other doctrines or independently, than peace of mind and a sure confidence, just as in all other branches of study. We must not try to force an impossible explanation, nor employ a method of enquiry like our reasoning either about the modes of life or with respect to the solution of other physical problems: witness such propositions as that 'the universe consists of bodies and the intangible', or that 'the elements are indivisible', and all such statements in circumstances where there is only one explanation which harmonizes with phenomena. For this is not so with the things above us: they admit of more than one cause of coming into being and more than one account of their nature which harmonizes with our sensations. For we must not conduct scientific investigation by means of empty assumptions and arbitrary principles, but follow the lead of phenomena: for our life has not now any place for irrational belief and groundless imaginings, but we must live free from trouble. Now all goes on without disturbance as far as regards each of those things which may be explained in several ways so as to harmonize with what we perceive, when one admits, as we are bound to do, probable theories about them. But when one accepts one theory and rejects another, which harmonizes just as well with the phenomenon, it is obvious that he altogether leaves the path of scientific enquiry and has recourse to myth. Now we can obtain indications of what happens above from some of the phenomena on earth: for we can observe how they come to pass, though we cannot observe the phenomena in the sky: for they may be produced in several ways" [183].

Further on in his letter to Pythocles Epicurus applies his principle of alternative explanations to a large number of astronomical and meteorological phenomena. The expression "scientific bankruptcy" would not seem too strong to describe this principle. Epicurus himself enlightens us about its purpose in a deliberate attack upon the astronomers, whose "devices"—i.e. their calculations and their progress in describing stellar movements—arouse his displeasure, because of the Platonic

167

association of perfect exactitude with the divine. "The size of sun and moon and the other stars is for us what it appears to be; and in reality it is either slightly greater than what we see or slightly less or the same size: for so too fires on earth when looked at from a distance seem to the senses. . . . The risings and settings of the sun, moon, and other heavenly bodies may be due to kindling and extinction. . . . Or again, the effect in question might be produced by their appearance over the top of the earth, and again the interposition of the earth in front of them: for once more nothing in phenomena is against it. . . . For all these and kindred explanations are not at variance with any clear-seen facts, if one always clings in such departments of enquiry to the possible and can refer each point to what is in agreement with phenomena without fearing the slavish artifices of the astronomers.

"The wanings of the moon and its subsequent waxings might be due to the revolution of its own body, or equally well to successive conformations of the atmosphere, or again to the interposition of other bodies; they may be accounted for in all the ways in which phenomena on earth invite us to such explanations of these phases; provided only one does not become enamoured of the method of the single cause and groundlessly put the others out of court, without having considered what it is possible for a man to observe and what is not, and desiring therefore to observe what is impossible. Next the moon may have her light from herself or from the sun. For on earth too we see many things shining with their own, and many with reflected light. Nor is any celestial phenomenon against these explanations, if one always remembers the method of manifold causes and investigates hypotheses and explanations consistent with them, and does not look to inconsistent notions and emphasize them without cause and so fall back in different ways on different occasions on the method of the single cause. . . . And do not let the divine nature be introduced at any point into these considerations but let it be preserved free from burdensome duties and in entire blessedness" [185].

The above quotations will suffice to show the detrimental influence of the philosophical axioms of the Epicurean School upon its scientific attitude. In fact, Epicurus here achieved the opposite of what he desired: he fostered the antithesis of heaven

and earth, though in the reverse sense, as it were, of that introduced by Plato. The means which he employed, both here and
in the problem of free will, were not such as to strengthen
scientific method as an independent discipline striving to attain
to methods of research and ways of comprehension prescribed by
its own needs alone. In respect to the law of causality and the
causal explanation of phenomena, the doctrine of Epicurus came
to grief through abandoning the Democritean tradition. It was
just here that the rival school of the Stoics distinguished itself by
an approach which in many respects is close to that of modern
science. The confusion of science with philosophy is, of course,
common to both schools. But, whereas the philosophy of Epicurus,
and especially his "religious complex", acted as a harmful
element in this confusion, the theological basis of Stoicism
actually assisted in the clarification of the problem of cause and
effect. We have seen how the notion of the divine logos as permeating the whole cosmos led the Stoics to conceive of the cosmos
as a continuum, all the parts of which are in dynamic interplay.
In this conception an important place was occupied, as we shall
now see, by causality. In contrast to the Epicurean theology,
which removed the gods from any "burdensome duties" into the
realms of eternal bliss where they had no responsibility for what
happened in the cosmos, the philosophy of the Stoics identified
the godhead with the supreme providence which watches over
everything at all times. In a way, this carries on Aristotle's teleological conception of "all is for the best"; but, at the same time,
the analogy between the continuity of providence and that of the
whole complex of cosmic events led to the identification of providence with the eternal chain of causality, with fate: uncompromising teleology joined hands with uncompromising determinism. Hence we observe the paradoxical situation that the
essentially religious Stoic School becomes the legitimate heir of
the Democritean conception of mechanical necessity and the
bitterest opponent of Epicurus' attempt to circumvent causality.
In the two following excerpts there is a clear reference to the
uncaused deviation of atoms postulated by Epicurus: "Chrysippus
confuted those who would impose lack of causality upon nature
by mentioning the die and the balance and many other things
which can never fall or swerve without some internal or external
cause. For there is no such thing as lack of cause and chance. In

the impulses mentioned, which some have arbitrarily called accidental, there are causes which are hidden from our sight and which determine the movement in a certain direction" [219]. Two points here deserve special notice: the appeal to experiment (the die and the balance), and the definition of chance as a hidden cause, i.e. as an expression of our inability to comprehend the full range of causality. On this subject we have succint evidence of the closeness of the Stoics to Democritus' view: "Anaxagoras, Democritus and the Stoics said that chance is a cause hidden from human comprehension" [111].

The second passage, which controverts the "deviation" of the atoms, contains the Stoic argument for the law of causality and merits special attention: "Everything that happens is followed by something else which depends upon it by causal necessity. Likewise, everything that happens is preceded by something with which it is causally connected. For nothing exists or has come into being in the cosmos without a cause. There is nothing in it that is completely divorced from all that went before. The cosmos will be disrupted and disintegrate into pieces and cease to be a unity functioning as a single orderly system, if any uncaused movement is introduced into it. Such a movement will be introduced, unless everything that exists and happens has a previous cause from which it of necessity follows. In their view, the lack of cause resembles a *creatio ex nihilo* and is just as impossible" [220]. The absolute rule of causality thus becomes an integral part of the cosmos conceived as a continuum, and the Stoic conception of the cosmos is completed by including the causal relation in it. This relation is precisely defined in our passage as a never-ending series of causes and effects held together by necessity. But especially important is the last sentence quoted, in which the validity of the causal law is compared with that of the laws of conservation (conservation of matter or of any other physical quantity). In Leucippus and Democritus we found a formula for each of these laws [70, 92]; but they are placed side by side as laws of the same kind for the first time in the teaching of the Stoics, where this analogy reveals a most profound grasp of the causality of nature which has been fully confirmed by modern physics. The mathematical formulation of physical laws, in itself proof of causality, has shown that certain of them can be formulated in the form of laws of conservation which express the fact that this or that physical quantity is conserved, or that there is a balance between certain quantities which means the permanence of physical conditions. We do not consider the possibility of formulating such laws of conservation as a coincidence; it is rather a further confirmation of causality, as can easily be shown by a simple example. According to our conceptions, the construction of a perpetuum mobile would invalidate not only the law of the conservation of energy but the entire law of causality, since to annul the fundamental laws of mechanics which find expression in the law of energy is tantamount to denying the possibility of describing the elementary phenomena of nature in terms of cause. The phrase "lack of cause resembles a *creatio ex nihilo*" is very apt, if it refers to the uncaused deviation of atoms from the straight line—for this deviation involves breaking the law of the conservation of momentum by the creation of "a quantity of movement" out of nothing.

The key term in the law of causality of Stoic physics is "fate", which even in the earliest Greek literature had served as an expression for necessity, and now became, especially in Chrysippus' works, a synonym for causality. Later sources have preserved several of his definitions, three of which are quoted from his book *On Fate* as follows: "Fate is the reason (logos) of the cosmos; or the reason of the events which occur in the cosmos under providence; or the reason for what has happened, is happening and will happen" [221]. The term "reason", indicating something rational, is found also in the sentence of Leucippus ("by reason and necessity" [70]) p. 159). Another formula particularly stresses the interconnection of things which is characterized by permanence and absence of deviation. "In his fourth book on Providence, Chrysippus says that fate is a certain physical order wherein one thing is always caused by and results from another, in such a way that this interrelation cannot be changed" [222]. The Stoics used the concept of fate to express an absolute, uncompromising determinism, in the sense given to it by classical physics, for instance in the famous formula of Laplace. Then, just as now, the thorough-going determinists included man and his works in the laws of fate; and at this point the Stoic School, like its rival the Epicurean, was confronted by the eternal problem of fate and free will. Since Stoic ethics were based upon man's responsibility for his deeds and on the belief in his power

171

to determine the course of his life, the Stoics could not give up free will. But could they maintain it without contradicting the very meaning of fate by some primitive device such as Epicurus'?

Of course, the Stoic solution is also unacceptable, but it is remarkable for its theoretical construction; and the physical analogy underlying it once more displays the analytical acumen of Chrysippus and his school as well as their great powers of scientific imagination. Chrysippus does not remove a single link of the causal chain; instead he divides causes into kinds— preliminary and determining causes. This division is influenced by the medical doctrines of his own day, echoes of which are still found in a later period. The preliminary cause is the impulse given to the disease by external conditions, while the determining cause is that which decides the course of the disease in accordance with the general physical characteristics of the patient. On this analogy, Stoic psychology distinguishes between the preliminary cause of human decision consisting of the external sensory stimulus, and the determining cause consisting of the person's innate qualities. "Chrysippus . . . distinguishes between various kinds of cause, in order to evade the necessity (of denying free will) while preserving fate. He says: 'There are primary and secondary causes. If we say that everything is determined by fate in accordance with preliminary causes, we do not mean the primary determining causes, but the secondary (subsidiary) ones. . . . Even though we have no control of the latter, we may still control our instincts'" [223]. Thus the Stoic theory is that man, as a creature endowed with intellect, can choose to let the stimulus act upon his nature or not. It is not our business here to trace the complications in which the Stoics became involved when they vainly tried to explain away the obvious difficulties arising from the differentiation of causes, and particularly from the fact that the exercise of this choice is also a link in the chain of preliminary causes which are subject to the laws of fate. But we should ponder for a moment the physical analogy which is cited in this connection: "If you throw a cylindrical stone down a slope, you are the cause of its descent by providing the impulse; but it will roll down not because of your activity, but because that is the nature of the stone and of the roundness of its shape" [224].

Here what starts the cylinder rolling is the preliminary cause, while the actual motion of rolling is determined by the determining cause. It is just by such a differentiation that theoretical physics to-day deals with problems of dynamics. To solve the actual problem of the motion of a physical system, the modern physicist puts into the equation of the motion all the "determining causes", that is to say, the qualities which belong to the system itself and the forces acting upon it. But this is not enough, since it gives us only an abstract, general solution; to discover the condition of the system at every instant of a concrete case, we have to determine its initial conditions, i.e. its position and velocity at a certain moment, at "zero hour". These conditions, together with the theoretical solution, provide a complete definition of the motion and apply the law of causality to the given case. The initial conditions are analagous to the "preliminary causes" in the Stoic doctrine of fate. By these conditions an artificial break is made in the infinitely ramified chain of cause and effect in order to put the problem in a form admitting of a practical solution. The initial conditions define a certain situation which is the result of causal developments and the details of which we either cannot, or need not, examine. It may be that the cylindrical stone in question was thrown down the slope by someone; or it may previously have rolled down the same slope; or it may have been broken off from a rock and fallen freely until it struck the slope. All that precedes the moment which interests us in the problem—the sum total of the "preliminary causes"—is included in the term "fate", no matter how complex the details may be. But this part of the story does not interest us; we take note only of its last link as a starting condition of the problem that we wish to solve. In his desire to have the best of both worlds, Chrysippus tried to preserve fate by means of "initial conditions" to which men are at every moment subject when they have to make a decision and which are outside their control; while at the same time he hoped to leave the door open for free will by postulating that their innate qualities are subject to their own control, in contrast to those of inorganic matter. This attempt led him to the penetrating analysis of how the causal law works in a concrete case, as expressed in the illustration of the stone on the slope.

This illustration is only one detail of a general picture which is revealed to us in Stoic physics perhaps more strikingly than in any other philosophical theory of Ancient Greece; we find a penetrating analysis of scientific method or of scientific reasoning which is either performed upon the wrong object or mixed up with worthless superstitions. Astrology and divination, which both were prominent in the physical teaching of the Stoics, undoubtedly came from the Orient and contain a strong element of irrationality. In connection with causality, it is interesting to note how the Stoics tried to put divination upon a rational basis. In the diviners' use of external signs and symbols the Stoics discovered the inductive method, whereby it is possible to foretell a future event through past experience of the connection in time between it and events which preceded it. "The results of those artificial means of divination, by means of entrails, lightnings, portents and astrology, have been the subject of observation for a long period of time. But in every field of enquiry great length of time employed in continued observation begets an extraordinary fund of knowledge, which may be acquired even without the intervention or inspiration of the gods, since repeated observation makes it clear what effect follows any given cause, and what sign precedes any given event" [225]. Here the principle of induction is employed in that very cautious form which Hume tried to take as the basis of the law of causality—the relation of events in time. This caution is not, of course, the same as an outright denial of the causal connection between those signs and the events that follow them; still, the question is obviously left open. Divination starts from the assumption that the constantly recurring sequence of the same signs and events can be expressed as a rule which justifies prophecy, whether it is really a causal law or simply a repetition of coincidences. "According to the Stoic doctrine, the gods are not directly responsible for every fissure in the liver or for every song of a bird; since, manifestly, that would not be seemly or proper in a god and furthermore is impossible. But in the beginning, the universe was so created that certain results would be preceded by certain signs, which are given sometimes by entrails and by birds, sometimes by lightnings, by portents and by stars, sometimes by dreams, and sometimes by utterances of persons in a frenzy. And these signs do not often deceive the persons who observe them properly. If prophecies, based on erroneous deductions and interpretation, turn out to be false, the fault is not chargeable to the signs but to the lack of skill in the interpreters" [226].

The previous quotation threw into relief the dependence of the law of induction upon experiment ("observation over a long period", "repeated observation"). This last reference, on the other hand, stresses the axiomatic aspect of the law resulting from that experiment, the belief that the universe is governed by law: "The universe was so created. . . ." In order to appreciate the full achievement of the Stoics in the question of induction and causality, we must for a moment forget all about the superstition in which divination was enveloped: what is important is that the Stoics here grasped a principle of vital importance in the process of understanding nature—the principle of mutual confirmation. "Chrysippus gives another proof in the abovementioned book of his: 'the predictions of the soothsayers could not be correct, if fate were not all-embracing'. . . . It seems that Chrysippus based his proof on the mutual interdependence of things. For he wants to show by the truth of divination that everything happens in accordance with fate; but he cannot prove the truth of divination without first assuming that everything happens in accordance with fate" [227]. The commentator here quoted did not at all understand that this criticism was really high praise of Chrysippus' intellectual grasp. The natural sciences are indeed based upon this mutual confirmation whereby every new instance of inductive evidence strengthens the law of causality, while the postulation of this law increases our confidence that any given chain of events is not arbitrary. Here Chrysippus came upon one of the fundamental categories of scientific epistemology. Characteristically enough, it was the anything but scientific signs of the soothsayers that gave rise to these epistemological conclusions. Still, it is well known that this is not the only time in the annals of science that correct conclusions have been drawn from false premises.

The idea of causality did not progress in the Ancient World beyond the Stoic conception of fate. The obvious reason for this was, first of all, the poverty of systematic experimentation. But it is clear that the religious significance given to fate, and the tendency to identify it with providence, impeded any further scientific development. On the other hand, there is an enormous advance from the first vague conceptions of a general conformity with law in nature to Democritus' conception of necessity, and from that to Chrysippus' specific grasp of the law of cause and effect. This development is particularly striking if we consider the law of causality as part of the general problem of the inter-connection of events and their dependence on time. This broader framework includes also the concept of probability and the aspect of the functional connection between phenomena. Therefore we should briefly survey the achievements in this field, too, if we wish to attain a proper appreciation of the intellectual struggles of the post-Aristotelian period to understand the causal network of things. From Pythagoras onwards, once the absolute certainty of a demonstrable mathematical statement had become part of men's intellectual outlook, the Greek word for "probable" was used, as opposed to mathematical certainty, in the sense of "likely, seemingly correct". "You do not produce any compulsive proof, but you make use of probability. If Theodorus and other geo-meters had employed this in geometry, they would be worthless" [123]. A similar note of contempt for inexactitude is heard in the following passage: "This assumption is, in my opinion, unproven and based upon what is probable and likely. For this reason it is accepted by many. But I know that arguments basing proofs on probability are like cheats, and, if not treated with care, may well mislead us, alike in geometry and in everything else" [120]. Plato, the author of these excerpts, understood that the whole of natural science is no more than probability. While for Plato this understanding led to a disdain of experimental science, Aristotle, though agreeing with this attitude in principle, stressed the practical value of investigation. "Plato rightly called natural science the doctrine of the probable. This is also the opinion of Aristotle, who declares that real proof is derived from sure, primary principles and from true, primary causes. But this does not mean that we should dismiss the natural sciences as worth-less, only that we should content ourselves with what is to our advantage and within our power, as Theophrastus too teaches" [172].

It was the progress of medicine and medical diagnosis, more than any other experimental science, that gained firm recogni-tion for the importance of the whole wide range of conclusions based on probability, where the guiding principle is not logical necessity but confidence in the regular recurrence of events through causal necessity. This recognition had, in its turn, a remarkable influence upon logic: in Aristotle's logic the first place is occupied by categorical statements and syllogisms, whereas his pupil Theophrastus began to develop the hypothetical syllogism. Subsequently, hypothetical and disjunctive sentences became the central feature of Stoic logic. These syllogisms, too, were used first of all in mathematics: in formal terms, the hypothetical syllogism ("if a, then b") is typical of the mathematical formula "if one of the angles of a triangle is a right angle, then the square on the hypotenuse is equal to the sum of the squares on the other two sides". The same is true of the disjunctive ("if a, then either b or c", etc.), as the following simple example shows: "If a whole number is divided by three, the remainder is either zero, or one, or two." Formally speaking, these syllogisms are applicable to everything empirical; and indeed, in the examples given in Stoic and subsequent literature, we can trace the gradual penetration of this kind of syllogism into the world of experience, whether it be everyday experience of ordinary natural phenomena, or the practice of the soothsayers or the foretelling of an individual's fate or of future historical developments. Especially striking in all this interesting development are the difficulties that came to light in the discussion of syllogisms containing the time factor ("if a, then subsequently b"; or, in the disjunctive syllogism, when various developments are possible: "if a, then b or c"). These difficulties, which run like a thread throughout Stoic literature from the earliest Stoics to the latest commentators, spring from the apparent incompatibility of the idea of possibility with necessity and fate. Fate means the clear-cut necessity by which things are held in mutual dependence; it means the absence of choice. Whereas the disjunction, as applied to experience, ramifies future development by distinguishing various possible occurrences. But is there really any possibility? Only one thing will actually occur; this then becomes the one necessary occurrence, and all the others are shown to be impossible. What is the place of the possible between the twin poles of the necessary and the impossible? The controversy in which the Stoics became involved with earlier philosophers was unfruitful. But it is worth while noting their characteristic endeavour to preserve the concept of possibility within the framework of their doctrine of fate, as we see it in the polemics between their interpreters and their opponents. "How can there not be a contradiction between the doctrine of possible occurrences and the doctrine of fate? If indeed, the category of the possible does not embrace either what is or will be true, as Diodorus postulates, but the term possible is to be applied to all that is likely to happen, though it never will happen; then there will be many possible things prevented from happening by the absolute and unassailable control of fate. Either the power of fate must dwindle, or what is likely to happen must become, in the majority of cases, impossible—if fate is really as Chrysippus supposed it to be. For all that exists is necessary, since it is part of the supreme necessity; and all that does not exist is impossible, since the most powerful cause prevents it from coming into existence" [228]. The position of the Stoics was defined still more clearly by another opponent: "There are those who would include the possible and likely as part of all that happens through fate, by defining the possible as something that is not prevented from occurring, even if it does not occur—'there is nothing to prevent the occurrence even of the opposite of what happens through fate, for even though it does not happen, it is still possible'. . . . Are not those who argue in this way just like jesters?" [229]. This limitation of possible occurrences to "those which are not prevented from occurring" is important: it shows that the Stoics were on the way to grasping the idea of probability as we, with our broader view of causality, understand it to-day. Instead of describing the *actual* occurrence by a series of steps which combine to form a chain in a single dimension, we describe the sum total of the *potential* occurrences —all the events that may happen within the framework of a given law or given combinations; and we describe them as a many-dimensional network of possibilities all of which are subject to the condition "that there is nothing to prevent their occurrence, even though they do not occur". In a game of dice, for example, the number of possible results from a throw is limited by the number of possible combinations; in the case of the excitation of the atoms of a gas in a discharge tube the complex of possibilities is determined by all the possible states of energy of the atom. In both these instances, every one of the possible occurrences has a specific degree of probability, which can sometimes be calculated or determined in another way. Since certainty exists only in the case of an event which has already occurred and belongs therefore to the past, we have grown accustomed to regarding the conformity to law of future events as a network of probabilities accessible to our investigation. We further consider the existence of this network and the possibility of expressing it in numerical terms as an expression of causality. In order to appreciate the importance of Chrysippus' attempt to fit possibility into the framework of fate, we should consider it in the light of this modern development of the concepts.

It is natural to ask at this point whether the Greeks arrived at any quantitative formulation of probability, even on the most elementary level, like their first steps in infinitesimal calculus. The disjunctive syllogism and the concept of possibility mark the starting-point of the calculation of probabilities. In this respect, the way was open for further theoretical progress. But not in this respect alone: there were also other circumstances favourable to such progress. As is well known, the development of the mathematical theory of probability in modern times, in the sixteenth and seventeenth centuries, was set in motion by the desire of card- and dice-players to know their prospects of winning. Now, we know that games of chance were very popular at all times in the Ancient World (indeed they have never ceased being played from the earliest times right down to the present). It is therefore reasonable to suppose that such games contributed to the development which we are discussing. But, as against this, we must note with astonishment that, for all the ubiquity and popularity of games of chance, they had no noticeable influence on scientific thought at any time in the Greek and Roman periods. We cannot discover any reference to the formation of the fundamental concepts of probability, such as expectation as the subjective aspect of probability, or the frequency of a constantly recurring event. Nor is there any mention of regularities appearing in random series (the law of large numbers), apart from the crudest formulations given by way of illustration. It may be argued that the milieu in which games of chance were played was not of such a cultural level as to affect the world of scientific thought, or even to gain mention in serious literature. But this was not so. On the contrary, such games were common in all classes of ancient society; many Roman Emperors were well known for their love of dice, and the Emperor Claudius actually published a book on the game. Still earlier, in Greece of the fifth and fourth centuries B.C., games of chance were customary in every circle: for example, Plato in his *Lysis* describes how Socrates entered the palaestrum on the Feast of Hermes and found the young men playing dice after the end of the sacrifices. The marking of the dice was the same then as it is to-day, viz. each surface had a different value from one to six, so arranged that the total of any two opposite surfaces was seven. A close second to the dice was the astragalos, made from the ankle-bones of a sheep, which is described in the second chapter of Aristotle's *Historia Animalium*. Unlike the symmetrical dice, the astragalos was an oblong such that it had only four surfaces of different sizes, one convex and its opposite concave, while of the two remainding, one was slightly sunken. The surfaces carried the numbers 1, 3, 4, 6, but the numerical value of a surface bore no relation to the frequency with which it would be thrown. Obviously, with this asymetrical natural object the chances of any given surface being thrown are various, whereas each surface of the dice has an equal chance—provided that the dice are not loaded. The absence of any relation between the markings customary on the astragalos and the frequency with which they were thrown points to lack of interest in the laws of chance. Further evidence in this direction can be found in the rules of the game, or, more precisely, in the absence of any connection between these rules and the simplest laws of probability. Four astragaloi were usually thrown together. Each of the thirty-five possible combinations was given a special name after some god or hero of mythology, and its value was not necessarily the same as the total number thrown. In some games the highest value belonged to the combination in which each astragalos showed a different number, e.g. 1, 3, 4, 6, even though this combination is far more frequent than a combination of the same numbers. According to other rules, the highest number won, i.e. the combination 6, 6, 6, 6. The most characteristic feature of the game is that there was no rule which made victory dependent on the result of a *series* of throws: in all the versions of the game, with all their different systems of reckoning, the game was won and lost on *one* throw by each of the players. This is yet another proof of lack of interest in the law underlying occurrences of this sort, so much so that the matter calls for investigation. It will be discussed in a later chapter. The games of chance which gave rise to the theoretical investigation of the laws of chance in modern times were just those which were won by the appearance of a certain number in each of a sequence of throws. This led people to study the relation borne by the *favourable events* to all the *possible* ones, in other words to the concept of expectation and everything connected with it. In the Ancient World, instead of calculating permutations, men concentrated on quickness of hand, on the skill of the player in throwing the desired number with the dice or astragalos; or, if no attempt was made to "woo Fortune", they trusted to blind chance or individual luck. "Nobody can become a skilled dice-player," says Plato in his *Republic*, "if he has not devoted himself to it from his childhood, but only plays for pleasure." (The context is the professional training of artisans.) Usually, however, the game of dice is quoted as the classic example of something unpredictable. At the end of the last book of Plato's *Laws* the Athenian stresses that all the details of the constitution cannot be thought out in advance, but that in many cases a procedure of trial and error must be followed: "With the constitution we must take the risk of throwing either three times six or three times one." Here the best possibility is compared to the maximum that can be thrown with three dice and the worst to the minimum. Aristotle too has an example in his *De Caelo* which resembles the one quoted from Plato in stressing the two extremes: "To succeed in many things, or many times, is difficult; for instance, to repeat the same throw ten thousand times with the dice would be impossible, whereas to make it once or twice is comparatively easy" [159]. This is apparently the most quantitative expression of an instance of the law of probability in classical literature. It confirms our general impression that a whole sphere of thought, which is one of the corner-stones of our view of life, was missing from the intellectual awareness of Greece and Rome.

Before concluding this chapter, some mention must be made of yet another fundamental type of interdependence. This is functional dependence, now an integral part of modern science, the first traces of which are found in Ancient Greece. In modern times, function has become a way of observing variables in terms of their interdependence. But the process really began with the mathematization of physics which followed upon the development of equations and analytical geometry, and particularly with the geometrical description of motion as a change of place functionally dependent on time. The observation of one class of variables in its relation to another which may vary as a result of its dependence on the first necessitates a certain dynamic approach. It further requires a more thorough understanding of the concept of a continuum, since the changes in the variables usually occur continuously. We should not be surprised, therefore, to find the first glimmerings of function in the Stoics, seeing that their cosmos was made up of dynamic elements, such as pneuma and tension, and was of a strictly continuous nature. The transition from constant to varying quantities can be seen already in Stoic logic, where, as noted above, great importance was attached to the disjunctive syllogism. In addition to the disjunction, which consists of a differentiation between distinct possibilities ("either day or night"), the Stoics further defined the comparative sentence which differentiates possibilities on an adjustable scale: "more day than night", or "more night than day". This is a most interesting variant of the usual disjunction and brings to mind the multi-valued logic of our own time. If we take the day as unity (full light=100 per cent. light), and the night as zero (lack of light=0 per cent. light), then the comparative sentence replaces the alternative "one or zero" by the adjustable scale of all the stages between zero and one. Thus the world of exact thought was enriched by one of the elements of the concept of function: the continuous variable. Another element—the relation between quantities—appears in the Stoic theory of categories. There relation is defined as one of the four principal categories. As an example, Stoic literature gives the concept of the father which is dependent upon his relation to the son, or the concept of the neighbour on the right which is dependent on the existence of a neighbour on the left. Chrysippus cites the vault, the stones of which are mutually supporting. This category leads him to define the quality of an object in terms of activity dependent upon its relation to another object. "As regards what happens in the air, the Stoics said that winter is the cooling of the air above the earth due to the withdrawal of the sun, while spring is the mellowing of the air due to the sun's return towards us. Summer is the heating of the air above the earth due to the sun's travelling northwards, while autumn comes about through the sun's turning away from us" [230]. Here the seasons of the year are clearly defined in terms of functional dependence—the dependence of climate upon the position of the sun. This position is here the independent variable, and its continuous changes bring about changes in the dependent variable, in the climate which determines the character of the season. This all shows that the concept of function came into being in the time of Archimedes and Eratosthenes. But this discovery was never followed up and never fructified ancient Greek science, which was then at its highest point. For this there are two reasons: lack of analytical geometry, and inability to comprehend time as an independent variable with phenomena as its function.

VIII

COSMOGONIES

"Shall a land be born in one day?"
IS. 66.8

———————

G REEK science appears to us as a continuous effort to
rationalize nature, resulting in the gradual extension of the
concept of law to every observed sphere of the physical universe.
We have traced the various aspects of this process and their
influence on the description of the physical phenomena of the
cosmos. But to complete this study we must add a survey of
ancient cosmogonies from the Ionian philosophers to the Stoics.
The term "cosmogony" means any description or explanation of
the creation of the cosmos, from the prescientific mythologies
found amongst all ancient peoples down to the scientific theories
of our own day which embody all our experimental and theoretical knowledge. Scientific cosmogony aims at describing the
formation of the universe by means of all the scientific laws and
data available at present. Since, by the nature of its subject, it is
obliged to apply its conclusions to very remote periods of the
world's history, it cannot be entirely free of the element of
speculation by which it was dominated in ancient times. In
cosmogony, more than in any other branch of science, we are
aware that the change from mythos to logos was not a sudden
jump but a continuous development: in many ancient mythologies indications can be found of an intentional rationalization
of the process of creation. In the two hundred years before the
beginnings of Greek philosophy, much of the ancient tradition

184

and legend about the coming into being of the gods and the cosmos found it way into Greek literature. Particularly noteworthy is the *Theogony* of Hesiod which had a profound and lasting influence on subsequent generations. The outstanding feature of all these mythologies is the personification of the forces of nature and their internecine conflicts, besides which the first chapter of Genesis appears remarkably scientific and rational. There the description of the six days of the Creation shows us the present orderly arrangement of the cosmos emerging from primitive chaos, as the word of the Creator, which serves as the supreme cause of this development, joins together the separate links into a natural sequence of events.

A characteristic example of the creation of order out of disorder is the process of separating out the opposites contained in the undifferentiated whole, as is strikingly described in Genesis 1: "And God divided the light from the darkness"; "And God divided the waters which were under the firmament from the waters which were above the firmament." This separation, which is simply the differentiation of unformed matter, is also the basic principle of the first scientific cosmogony on which all subsequent cosmogonies were modelled. Its author was Anaximander: "He says that that which is capable of begetting the hot and the cold out of the eternal was separated off during the coming into being of our world, and from this there was produced a sort of sphere of flame which grew round the air about the earth as the bark round a tree; then this sphere was torn off and became enclosed in certain circles or rings, and thus were formed the sun, the moon, and stars" [6]. We have seen that the primeval matter of Anaximander is the Infinite which cannot be qualitatively defined. Thus, according to him, the creation of the world began when a certain portion of this formless mass was detached from the rest, thus setting in motion a process of differentiation which produced the beginnings of order by separating out the two opposed qualities, hot and cold. The significance of this separation is twofold. First of all it means, in the terminology of modern physics, that every physical occurrence in the cosmos can only come about through the existence of a difference in potentials which makes a transition from one level to another possible, e.g. thermic differences, differences in the gravitational or electric potential, etc. For this reason, the task of every cosmogony, from

Anaximander to the present day, is to explain how such gradients came about in a homogeneous environment which does not contain them. Secondly, as regards the particular case of the specific contrast picked on by Anaximander, our cosmos displays the special separation of cold, which is located on the earth, from hot, which is found in the heavens: the way in which this division came into being also calls for a physical explanation.

What was the cause of the original separation? Various doxographic sources provide the answer to this question: "He (Anaximander) explains the creation of the world not by any change of the primeval matter, but the separation of opposites as a result of the eternal motion" [5]. This shows us that the earliest scientific cosmogony was of a purely mechanical nature and that Anaximander's aim was to explain the formation of the cosmos on physical principles, admitting only natural causes, in the same rationalistic way as that in which he and his followers had explained the phenomena around them. The nature of this eternal motion is clear from other sources, as also from the cosmogony of the atomic school which was constructed on Anaximander's principles. The model employed (which we shall examine in more detail below) was the circular movement of a vortex of water or air. The primeval matter was conceived as revolving in such a vortex by virtue of some inherent quality. Neither Anaximander nor the atomists ask how this motion came into being: it exists from all eternity and is inseparable from matter. Similarly, when discussing the atomic theory, we saw that Democritus and his followers accepted the movement of the atoms as an ultimate, inexplicable quality which had to be assumed for the description of visible nature. In the case of the cosmogonic explanation, however, there is an important difference. The movement of the atoms in the void was depicted as observing no order or fixed direction, and the various combinations of atoms, i.e. the macroscopic bodies, were produced in accordance with the laws of statistics. The movement of the vortex, on the other hand, is not devoid of order: it is a revolution in one direction of all the primeval matter. Thus, the assumption of its existing from the beginning involves abandoning an important aspect of the general hypothesis of a primitive state of complete chaos when the world came into being. This explains Aristotle's criticism implied in the following sentences: "There are some too who ascribe this heavenly sphere and all the worlds to spontaneity. They say that the vortex arose spontaneously, i.e. the motion that separated and arranged in its present order all that exists" [113]. Of course, Aristotle goes further, and from his teleological standpoint rules out any possibility of order resulting from circular motion in obedience to mechanical laws. But, even without accepting his teleological approach, we may regard his strictures about the formation of this "spontaneous" movement as pertinent.

The separation of hot from cold by the vortical movement led, according to Anaximander, to the concentration of the cold element in the centre and of the hot at the periphery of the vortex in the form of a sphere of fire. Between these opposites stretched a sort of no-man's-land of air. Through the momentum of the vortex, the sphere of fire was broken up into a series of wheels which were enveloped in air and mist. The heavenly bodies were openings in the envelopes, the revolutions of which still reflected the circular movement of the original vortex. The cold centre was composed of the two "cold elements" in combination—earth and water. Through the action of heat, part of the moisture evaporated, while the rest of the water combined to form the ocean, which too is still evaporating. From the doxographic sources it appears probable that Anaximander assumed that there was an infinite number of universes existing simultaneously (cf. quotation 6 on p. 9), and that every one of them was created by the process which began with the separation of a unit of primordial matter from the body of the infinite. "He declared that destruction and, long before this, generation has been going on from infinitely distant ages, all the worlds recurring in cycles" [6]. We are not told how Anaximander pictured the details of the transition from order to disintegration; nor do we know how this method of giving a natural explanation of creation led him to conceive the opposite of this process. This question may just as obviously be asked about the process of creation; since the cosmos contains no clear evidence of its formation or destruction, it would have been more plausible to assume that it is eternal. Possibly the creation and destruction of worlds in cycles which we find in the ancient mythologies indicate that the periodicity of nature, especially with regard to the seasons of the year, was given a universal significance and projected on to the cosmic plane. Thus the cosmos itself was subject to the eternal cycle of germination, flowering and decay which is revealed in miniature, so to speak, in the terrestrial cycle of seasons and in other meteorological phenomena. Aristotle again and again emphasized the universality of this process: "Now we observe in Nature a certain kind of circular process of coming-to-be. . . . In actual fact it is exemplified thus: when the earth had been moistened an exhalation was bound to rise, and when an exhalation had risen cloud was bound to form, and from the formation of cloud rain necessarily resulted, and by the fall of rain the earth was necessarily moistened: but this was the starting-point, so that a circle is completed" [140]. This "small" cycle no doubt served as a model for the cyclical formation and destruction of the whole cosmos. At the end of the cycle the cosmos will return to the absolute equalization of all the opposites and elements in it, as is constantly occurring on the smaller scale of meteorological phenomena. Further evidence for this analogy is found in the doctrine of Heracleitus, who said that the moisture which rises from the rivers and seas provides the sun and stars with fuel. In this way, fire was included in the cycle of the elements. The doctrine of Heracleitus was taken up by the Stoics, and we shall return to it later.

From Aristotle we learn why the vortex was chosen as the model, "If the earth's rest is due to constraint, it must have been under the action of the vortex that it travelled into the middle. This is the name which all agree in giving to the cause, reasoning from what happens in liquids and in the air, where larger and heavier things always move towards the middle of a vortex. Therefore all who hold that the world had a beginning, say that the earth travelled to the middle" [66]. Vortical movements in the air, circular wind-squalls, are common phenomena. The physical details of this aerodynamic process are complicated. Many factors are involved, including the friction at the bottom of the vortex, which is a friction between air and earth. The movement of a dust spiral, or of fallen leaves gyrating in the wind, shows the direction of the velocity and its size at the various points of the vortex. Light dust is raised by the wind and swept along in the general rotatory motion, whereas heavier bodies which the wind is hardly strong enough to lift remain where they are in the centre of the vortex. They are joined by others which are close to the centre and are drawn towards it in constantly decreasing circles by a frictional motion. Such a vortex was an extremely convenient and vivid model of the way in which the geocentric cosmos was thought to have come into existence. It enabled the first cosmologists to indicate the mechanism whereby the heavy material was piled up in the centre, while the light and rare rotated around it. The process of separation could thus be explained as resulting from the momentum of the vortex.

The significance of rotatory movement for any cosmogonic explanation has been given added prominence in modern times by the gradual evidence of its universality in the cosmos. Galileo discovered the rotation of the sun on its axis; and the perfection of the telescope has revealed the axial rotations of the planets which Plato had assumed *a priori*. The planets revolve around the sun in one and the same sense; and double stars revolving around their common centre of gravity are a frequent phenomenon in our galaxy. The closer study of the highest units of the cosmos, the extragalactic nebulae, has left no doubt that rotatory movement is found in their mighty systems. In view of the structure of galaxies with their arms stretching out from a dense centre and curling round it spirally, it is hard to resist the conclusion that their rotation is of the vortical kind. Thus it is that the latest cosmogonies, which aim at explaining the structure of the cosmos and the formation of the galaxies, have returned to the vortex model and apply the laws of hydrodynamics to problems of cosmogony. The laws of mechanics tell us that for every given quantity of rotatory movement occurring in a system there must be a quantity of equal size and opposite sense in the same system or another one which has been in contact with it. In other words: rotation cannot be created out of nothing. Hence, if there are many such systems, it may be assumed that the different senses of rotation are statistically distributed over them. Whatever the mechanism by which the rotations emerged from the primeval state of chaos which preceded the creation of the universe, it is clear that statistical disorder must have prevailed with regard to the rotatory movement of the whole cosmos too. But with regard to a system which is, from the start, isolated from every other one, or with regard to the cosmos of the Greek cosmologists which was composed of the earth in the centre with the heavens around it, Aristotle's question is most pertinent: How was the vortex formed "spontaneously"?

The first serious modern attempt at a cosmogony, Kant's theory, foundered on the same difficulty as Anaximander. In his essay *Allgemeine Naturgeschichte und Theorie des Himmels* (1755), Kant tried to give a mechanical explanation, based on Newton's law of gravitation, for the formation of the solar system. One of the most striking features of this system is the uniform sense of its rotations (with a few negligible exceptions) alike in the movements of the sun and planets on their axis, the revolutions of the planets round the sun and the revolutions of their own satellites. Kant assumed that the mass of the whole system was scattered throughout space at the beginnings of the creation and that the process began with the gravitational movement of the particles to a place where an excessive density of matter chanced to occur. Being unaware of the law of conservation of rotatory momentum, Kant supposed that from this anarchic condition of particles streaming in disorder from all sides in one direction and at the same time colliding with each other, there spontaneously emerged a uniform rotation in one and the same sense. Laplace and subsequent cosmologists avoided Kant's mistake by assuming that a state of slow rotation already existed in the original mass out of which the solar system developed. This rotation was transmitted to all the parts of the system separated off from it. About the cause of this original rotation Laplace was silent, since it was clear to him that it must lie outside the solar system.

Anaxagoras was possibly the first to perceive the great difficulty involved in the creation of the vortex from nothing, as suggested by Anaximander, and to propose a solution of the problem. In his doctrine we find Mind as the force which brings order into the cosmos. The term "mind" suggests something non-physical; and indeed from the extant fragments of Anaxagoras it appears that Mind is distinct from matter and is not a compound made up from "the seeds of matter", as he called them. But the juxtaposition of mind and the seeds of matter makes us inclined to think that mind, too, is essentially physical. It would thus seem that Anaxagoras propounded a sort of monistic modification of Empedocles' forces of love and hate, with mind as a kind of force

190

and one of its important functions the setting in motion of the first vortex which marked the start of the creation: "Mind took command of the universal revolution, so as to make (things) revolve at the outset. And at first things began to revolve from some small point, but now the revolution extends over a greater area and will spread even further. And the things which were mixed together, and separated off, and divided, were all understood by Mind. And whatever they were going to be, and whatever things were then in existence that are not now, and all things that now exist and whatever shall exist—all were arranged by Mind, as also the revolution now followed by the stars, the sun and moon, and the Air and Aether which were separated off. It was this revolution which caused the separation off. And dense separates from rare, and hot from cold, and bright from dark, and dry from wet" [57]. From this, and from Aristotle's criticism, it seems that the Mind of Anaxagoras acts as a natural cause within the framework of the laws of nature, and has nothing to do with the teleological purpose of Aristotle's cosmos: "Anaxagoras uses reason as a *deus ex machina* for the making of the world, and when he is at loss to tell from what cause something necessarily is, then he drags reason in, but in all other cases ascribes events to anything rather than to reason" [67]. Most interesting is Anaxagoras' hypothesis that the rotatory movement began in one small point from which it spread outwards. Unfortunately, no further details of this hypothesis have come down to us. From the rest of his fragments we obtain a picture similar to the cosmogony of Anaximander: "And when Mind began the motion, there was a separating-off from all that was being moved; and all that Mind set in motion was separated; and as things were moving and separating off, the revolution greatly increased this separation" [58]. "From these, while they are separating off, Earth solidifies; for from the clouds, water is separated off, and from the water, earth, and from the earth, stones are solidified by the cold; and these rush outward rather than the water" [59].

According to another fragment, the momentum of the rotation constantly increased until it exceeded all terrestrial velocities, as shown by the movement of the heavens. The sentence in [59] that the stones were forced outwards more than the water, agrees with what we have learnt of Anaxagoras' theory about the nature

191

of the sun and stars: for him they were stones that had become fiery through the velocity of their rotation [62, 63].

Anaximander's attempt to explain the formation of the cosmos by a natural mechanism was taken over into the atomic school, which adapted the fundamental ideas of his cosmogony to its own basic principles. The atomists replaced the continuous unqualified primeval matter, which preceded the creation, by the variously shaped atoms which were jumbled together throughout infinite space. Here, too, it is the cosmic vortex that reduces chaos to order, this time, however, not by the separating out of opposites but by a sifting process which collects different kinds of atoms together one on top of the other until the conglomerates become macroscopic bodies with definite characteristics. The closest parallel to this process is the shaking of a sieve to and fro so that it separates out grains according to their size, or any other mechanical activity which brings about the separation of homogeneous particles. A very concrete illustration is found in a fragment of Democritus: "Living creatures consort with their kind, as doves with doves, and cranes with cranes, and similarly with the rest of the animal world. So it is with inanimate things, as one can see with the sieving of seeds and with the pebbles on beaches. In the former, through the circulation of the sieve, beans are separated and ranged with beans, barley-grains with barley, and wheat with wheat; in the latter, with the motion of the waves, oval pebbles are driven to the same place as oval, and round to round, as if the similarity in these things had a sort of power over them which had brought them together" [90].

The cosmogony of Leucippus, the first of the atomic cosmogonies, contains many details which, in the form in which they have come down to us, are not clear. Its lineal descent from the cosmogony of Anaximander is especially felt in the idea of the separation of one unit from an infinite mass, and likewise in the absence of any hint at the origin of the vortex. Although the atomists postulated movement as an inherent quality of the atom, like its shape, the essential question, as we have already seen, still remains: what explanation can be given for the creation of an orderly rotatory motion in one sense out of the sum of disorderly and undirected movements? "The worlds come into being thus. There were borne along by 'abscission from the infinite' many bodies of all sorts or figures 'into a mighty void', and they being gathered together produce a single vortex. In it, as they came into collision with one another and were whirled round in all manner of ways, those which were alike were separated apart and came to their likes. But, as they were no longer able to revolve in equilibrium owing to their multitude, those of them that were fine went out to the external void, as if passed through a sieve; the rest stayed together, ·and becoming entangled with one another, ran down together, and made a first spherical structure. This was in substance like a membrane or skin containing in itself all kinds of bodies. And, as these bodies were borne round in a vortex, in virtue of the resistance of the middle, the surrounding membrane became thin, as the contiguous bodies kept flowing together from contact with the vortex. And in this way the earth came into being, those things which had been borne towards the middle abiding there. Moreover, the containing membrane was increased by the further separating out of bodies from outside; and, being itself carried round in a vortex, it further got possession of all with which it had come in contact. Some of these becoming entangled, produced a structure, which was at first moist and muddy; but, when they had been dried and were revolving along with the vortex of the whole, they were then ignited and produced the substance of the heavenly bodies. The circle of the sun is the outermost, that of the moon is nearest to the earth, and those of the others are between these. And all the heavenly bodies are ignited because of the swiftness of their motion; while the sun is also ignited by the stars. But the moon only receives a small portion of fire" [83]. Here too we find the same arrangement in the separation of the elements, whereby the fine particles are placed on the perimeter, while the coarse are concentrated in the centre. The skin envelop expanded and created the heavens in whose vastness the agglomerations of fine particles blazed forth into stars, while the mass in the centre solidified into the earth's globe.

Although no such detailed record of Democritus' cosmogony has come down to us, the few references found in Diogenes point to a change of view in relation to Leucippus: "The atoms . . . are carried round by a vortex in the void and so all the compound bodies—fire, water, air and earth—are formed. They are structures made of atoms. . . . The sun and the moon are composed of smooth, round atoms, like the soul. . . . And everything is created in accordance with necessity; for the vortex is the cause of the creation of everything and he calls it necessity" [112]. From this it would appear that Democritus identified the fine atoms which were thrown out to the perimeter with the "smooth, round" atoms of fire. In that case, it was not the velocity of the rotation that ignited the heavenly bodies, as Leucippus said following Anaxagoras, but the actual agglomeration of the light atoms of fire in one place. Another source, however, gives a different picture and makes it appear that, in Democritus' view, the sun was first created as the dark centre of a particular cosmos and subsequently caught fire through absorbing atoms of fire: "He speaks of the creation of the sun and the moon. These used to move independently, not having anything of the nature of heat and without any brightness; on the contrary, they had a nature like that of the earth. For each of them was first formed as a foundation of a cosmos of its own. Subsequently, however, as the sphere of the sun grew larger, the fire was entrapped inside it" [94]. The interesting point in the first of these two passages is the emphasis laid on necessity, i.e. the law of causality, in connection with the vortex. Democritus, though apparently unable to explain the details of its formation from the general movement of the atoms, understood the great importance of this point and therefore reiterated that "everything is created in accordance with necessity" and that the vortex itself, the cause of everything, is simply necessity. He was obliged to take up this position by his refusal to admit the existence of forces and his insistence that the movements and impacts of the atoms were the basis of everything. Even though he could not see any mechanism by which the vortex could have been brought into being, he believed that such a one must exist in accordance with the laws of causality. Possibly this was meant as an implied criticism of the marshalling Mind of Anaxagoras.

Another integral part of the atomic cosmogony was the belief in a vast number of worlds in a constant process of being created and destroyed. With regard to the destruction of worlds, Democritus hit upon a new idea: order is destroyed by the collision of one cosmos with another. This thought is expressed in the following text, which is remarkable in many respects: "There are worlds infinite in number and different in size. In some there is neither sun nor moon, in others the sun and moon are greater than with us, in others there are more than one sun and moon. The distances between the worlds are unequal, in some directions there are more of them, in some, fewer; some are growing, others are at their prime, and others again declining, in one direction they are coming into being, in another they are ceasing to be. Their destruction comes about through collision with one another. Some worlds are destitute of animal and plant life and of all moisture. In our world, the earth came into being before the stars; the moon has the lowest place, then the sun, and after them the fixed stars. The planets themselves are not at equal heights. A world continues to be in its prime only until it becomes incapable of taking to it anything from without" [114]. The atomists' reason for limiting the cosmos in time is clearly given by Lucretius: "Since the elements of which we see the world composed . . . all consist of bodies that are neither birthless nor deathless, we must believe the same of the world as a whole. . . . So, when we see the main component members of the world disintegrated and reborn, it is a fair inference that sky and earth too had their birthday and will have their day of doom" [253].

Of Epicurus, who carried on the teachings of Leucippus and Democritus, we have already spoken above. We saw that, while his great contribution was the working out of the details of the atomic theory, he displayed no special originality in the elaboration of its foundations. His most important contribution was the coining of the concept "association of atoms", i.e. the molecule. His cosmogony is included in a letter to Pythocles: "A world is a circumscribed portion of sky, containing heavenly bodies and an earth and all the heavenly phenomena whose dissolution will cause all within it to fall into confusion: it is a piece cut off from the infinite and ends in a boundary either rare or dense, either revolving or stationary: its outline may be spherical or three-cornered, or any kind of shape. . . . And that such worlds are infinite in number we can be sure, and also that such a world may come into being both inside another world and in an interworld, by which we mean the space between worlds; it will be in a place with much void, and not in a large empty space quite void, as some say: this occurs when seeds of the right kind have rushed in from a single world or interworld, or from several: little by little they make junctions and articulations, and cause changes of position to another place, as it may happen, and produce irrigations of the appropriate matter until the period of completion and stability, which lasts as long as the underlying foundations are capable of receiving additions. For it is not merely necessary for a gathering of atoms to take place, nor indeed for a whirl and nothing more to be set in motion, as is supposed, by necessity, in an empty space in which it is possible for a world to come into being, nor can the world go on increasing until it collides with another world, as one of the so-called physical philosophers says. For this is a contradiction of phenomena" [184]. In this cosmogony Epicurus takes Leucippus and Democritus to task. There is some truth in the first part of his criticism that "such a world may come into being in a place with much void and not in a large empty space quite void". This remark was directed against the traces of Anaximander's conception found in Leucippus' description of a unit detached from the Unlimited and transferred to a vacuum in which the cosmos was then formed. For infinite space is, after all, a mixture of emptiness and atoms outside of which there is no place for the birth of the cosmos.

But the question still remains: how were our orderly cosmos and all the other infinite number of worlds, the existence of which Epicurus also concedes, created of this confused mass of atoms? Epicurus does not find in the model of the vortex an adequate description of the beginning of the movement: in his opinion we have to explain in detail how the movement arose out of the previous situation and not take refuge in the mere term "necessity", as Democritus had done. After this criticism, we wait in vain for Epicurus himself to make good the deficiency; on the contrary, he only makes confusion worse confounded. He speaks of the creation of one world from atoms which have escaped from another or which have come from space and accumulated between the worlds. Does he suppose that these atoms flow in a straight line and with uniform velocity, as in the description found in Lucretius (cf. [248] on p. 163)? And did the cosmic condensation begin with that famous deviation from the straight line which introduced free will into the creation? Epicurus makes no mention of all this in connection with his cosmogony. Even if we suppose this to be the case, how did he picture the subsequent stages of the creation? Perhaps he conceived of them in the form of a great number of small rotatory movements which eventually coalesce into the vortex. Or perhaps he considered the single vortex superfluous and imagined that all the heavenly bodies and the earth were formed out of many small vortices. The parallel description of Lucretius throws no fresh light on this question: "Certainly the atoms did not post themselves purposefully in due order by an act of intelligence, nor did they stipulate what movements each should perform. But multitudinous atoms, swept along in multitudinous courses through infinite time by mutual clashes and their own weight, have come together in every possible way and realized everything that could be formed by their combinations. So it comes about that a voyage of immense duration, in which they have experienced every variety of movement and conjunction, has at length brought together those whose sudden encounter normally forms the starting-point of substantial fabrics—earth and sea and sky and the races of living creatures" [254]. One thing, however, is clear: Epicurus (and Lucretius after him) emphasized the importance of statistical disorder at the beginning of the creation and believed, like Kant centuries later, that from this complete disorder there would "somehow" emerge a state of order, i.e. orderly rotations. After all, an eternity of time is allowed for the creation and it is therefore inevitable by the laws of chance that at some time there should occur combinations capable of developing into a constantly extending orderly system. In principle, this adds nothing to Democritus except a pictorial representation of the workings of his "necessity". It may well be, therefore, that Epicurus in his description took for granted the place and function of the uncaused deviation which was an integral part of his philosophy.

However that may be, the cosmogony of Epicurus contains another idea which is an original contribution of real scientific significance. He sees infinite space as full of worlds of various shapes all of which are formed in accordance with the same mechanical laws. These worlds increase in size through supply from the infinite atomic material which is spread very thinly throughout the spaces between them. This increase has an upper limit fixed by the individual laws of each cosmos. Epicurus rejects the view held by Democritus that a cosmos could go on expanding until it collided with one of its neighbours. He maintains that the cosmos grows "as long as the underlying foundations are capable of receiving additions", but, once the saturation point is reached it proceeds to disintegrate. Here let us recall Epicurus' conception of molecules: he pictured them as associations of atoms having a defined structure and moving through the void, just like the atoms themselves, and capable of attracting any isolated atoms in their vicinity. At the same time they preserve their own particular structure through the vibrations of the atoms composing them. This suggests an interesting explanation of Epicurus' picture of the cosmos. He regarded every cosmos as a kind of enormous molecule with a well-defined structure and in a constant state of exchange with the atoms outside it, the laws of its growth and stability being determined by the particular dynamics that hold its parts together. Here again Epicurus displays that power of scientific inference and creative imagination which enabled him to transfer basic concepts and models from the dimension of the atom to the dimension of the cosmos. We are reminded of the cosmogonies of our own time which likewise regard agglomerations of stars as atoms of a gas, or explain the dynamics of nebulae by the laws of hydrodynamics, the guiding principle being that under certain conditions laws ascertained for smaller entities might be applied to entities of larger dimensions.

The conception of the formation and decay of the cosmos was given a different twist in Stoic doctrine. The views of the Stoic School were considerably influenced by the philosophy of Heracleitus (beginning of the fifth century B.C.), who, like Empedocles after him, regarded the harmony prevailing in the universe as the result of a dynamic equilibrium of opposite forces. For Heracleitus, this dynamics was built up round fire: "There is an exchange: all things for Fire and Fire for all things, like goods for gold and gold for goods" [37]. The sun and stars were created from, and are still fed by, the evaporation of the water on the surface of the earth; and this evaporation is brought about by the heat reaching the earth from the heavenly bodies. This double motion, upwards and downwards, is characteristic of the harmony of opposites whereby there is a simultaneous process of coming into being and decay within the existing cosmos. This was Heracleitus' first theory, as confirmed by the extant fragments of his works, and we should understand the following passage from Aristotle accordingly: "All thinkers agree that it has a beginning, but some maintain that having begun it is everlasting, others that it is perishable like any other formation of nature, and others again that it alternates, being at one time as it is now, and at another time changing and perishing, and that this process continues unremittingly" [39]. The notion of the eternal order of the cosmos as maintained by a simultaneous process of creation and decay was, in the generations after Heracleitus, mixed up with the idea of creation and destruction occurring one after the other, and with cyclical processes in which fire still played the essential part: the cosmos develops out of fire and eventually returns to fire, and so on in endless cycle. This idea appears in ancient mythologies. But the Stoics incorporated it into their scientific doctrine as a modification of Heracleitus' theory and set him up as the authority for their conception of the final conflagration of the cosmos. It is, therefore, hardly surprising that later commentators and compilers completed the confusion by identifying the theory of the Stoics with that of Heracleitus: "Heracleitus too says that the cosmos goes up in fire and is once again formed out of fire at certain periods of time in which, according to him, it is 'kindled in measure and quenched in measure'. This opinion was later arrived at by the Stoics too" [40].

We have seen that in Stoic physics fire occupied a special position amongst the elements by virtue of its active character. The Stoics were the first to grasp the special significance of thermodynamic processes in inorganic nature, as well as in biology, where their great importance had already been recognized before. It is thus easy to understand why the Stoic cosmogony was also built upon the concept of fire as the symbol of thermic phenomena: "Zeno said that fire is the essence of what exists. . . . At periods of time allocated by fate all the cosmos is conflagrated and after that it returns to its first order" [190]. This doctrine of the founder of the school was carried on by his followers: "Zeno, Cleanthes and Chrysippus hold that matter undergoes transmutation, as e.g. fire turns into seed, from which once again is restored the same world order, that was before" [191]. The process of the emergence of the cosmos from the primary state pictured as primeval fire is essentially the differentiation of the other three elements from fire: "Chrysippus, in his first book *On Nature*, says: The transmutation of fire is as follows. It turns first into air, then into water; and from the water at the bottom of which earth settles, air rises. As the air becomes rarefied, the aether is spread around in a circle; and the stars and the sun are kindled from the sea" [231]. This is the first stage of the creation in which the nature of the cosmic continuum is defined principally by the hot and dry elements. The emergence of air from fire as a second element is pictured on the analogy of the ascent of smoke and hot air from flames. This air (including all kinds of vapour) then condenses into water. From here onwards the development proceeds in two different directions: downwards for the sedimentation of earth from water, and upwards for the evaporation of vapours which, through rarefication, become fire, and this fire then agglomerates into the heavenly bodies. This completes the creation of the physical universe, which continues to maintain itself in being by the dynamic equilibrium of Heracleitus. However, it appeared that this equilibrium between the water in the seas and the fire in the sun is only a first approximation to reality: on a cosmic scale of time fire holds the ascendancy. Evidence for this was found in geological indications (already remarked on by pre-Stoic natural philosophers), such as the presence of shells on dry land showing that the sea has retreated, and in slow meteorological changes over the course of centuries: "Swamps and wet places become habitable through dryness, while places that were previously habitable become uninhabitable through the increase of dryness. This is explained by the change and decay of the universe. From these signs there are some, like Heracleitus and his followers, together with the Stoics, who believe that eventually there will be a conflagration of the universe" [232].

It follows from this that we are not to regard the conflagration of the cosmos as an actual fire, a kind of holocaust which in a moment destroys the whole cosmos: the process, in fact, is an extremely slow one, like all cosmic processes, the rate of decay being equal to that of coming into being. In the end, the original situation will be restored and the active element will dominate all the vast expanses of the cosmos which will then be reborn from the primeval fire. With the "thermic" cosmogony of the Stoics before us, we can hardly fail to be reminded of the "thermic death" of the universe. This idea formed the subject of many controversies amongst nineteenth-century physicists, when the second law of thermodynamics was applied to the universe as a whole. Every physical process ultimately leads to an increase in thermic energy. Thus, after the equalization of temperatures, the universe will reach a state similar to that of the final state of the Stoic cosmos.

Since the human mind is not unreasonably appalled by this prospect of an absolute, irrevocable end, various physical solutions out of the dilemma have been proposed to enable the universe to pass safely over the dead point of its final doom. Statistical mechanics explains the thermal phenomena as kinetic energies of the atoms and molecules. In the final state of the universe the movement of all these elementary particles will be one of ideal disorder and their velocities will be equal, on the statistical average. This is where statistics step in and save the situation: through a great fluctuation, a deviation from the statistical mean is likely to occur at some time, as the law of large numbers tells us. A great number of particles with a velocity considerably above the average will then agglomerate in one place and the resulting difference in potentials will bring about a renewal of the life of the cosmos.

Thus statistical mechanics offered a possibility of endless cycles of cosmic creation and destruction in a way that is closely parallel to the Stoic conception. As a corollary of this statistical picture of the cosmos, the old problem of the "return of the identical" came to life again in the philosophical debates of the last century. If the state of the cosmos is defined by a certain combination of its ultimate particles, and if every new combination necessarily results, by the laws of causality, from its predecessor, it follows that eventually, after all the permutations have been exhausted, the whole cycle of previous combinations will be repeated. Hence the idea of cosmic cycles, when comprehended in terms of number, involves the identical repetition of the present, a fact which was seized on by philosophers like Nietzsche for the purpose of their own doctrine. It is most interesting to observe that the very same inner logic which in our times has linked together thermodynamics, statistical mechanics and the idea of the "return of the identical", was also the driving force in Greek thought. Then, too, within the limits of the scientific comprehension and terminology of the time, views on the cosmological significance of thermal processes were closely connected with the theories about the formation and destruction of the cosmos and about the identical return of a situation. Of the Stoics we are told that "in their view, after the conflagration of the cosmos everything will again come to be in numerical order, until every specific quality too will return to its original state, just as it was before and came to be in that cosmos. This is what Chrysippus says in his book *On the Cosmos*" [233]. An even clearer statement is found in another source which quotes Chrysippus' own words from his book *On Providence*: "Clearly, it is not impossible that, after our death, when long aeons of time have passed, we shall return to the form we have at present" [234]. From Eudemus of Rhodes, a pupil of Aristotle's and the first to write a history of astronomy in the Ancient World, we learn that Pythagoras had anticipated the Stoics' conception of the return of the identical. In the following passage, Eudemus addresses his pupils during his lecture: "It may be asked if the same time will return, as some say, or not. . . . If you believe the Pythagoreans everything will eventually return in the self-same numerical order and I shall converse with you staff in hand and you will sit as you are sitting now, and so it will be in every-thing else. It is reasonable to assume that time too will be the same. For movement is one and the same, and likewise the sequence of many things that repeat themselves is one and the same, and this applies also to their number. Hence everything will be identical, including time" [36].

We see, then, that the Pythagoreans, like the Stoics after them, also tried to put the theory of cosmic cycles on a scientific basis, in their case in terms of their theory of number. Even though they had no knowledge of the theory of permutations nor any conception of the law of large numbers, the Pythagoreans realized that number by its very nature includes a certain order determined by law. The periodicity of cyclical motion symbolizes the periodicity of every cosmic event and therefore that of time itself. We have already seen in another Stoic fragment [233] that the Stoics also conceived of this Pythagorean idea of repetition "in numerical order"; they completed the physical picture of their cosmogony by the use of mathematical terminology.

The Stoics, uncompromising as they were in their conception of the physical world as a continuum, supposed that "beyond the cosmos there stretches an infinite, non-physical void" [206]. This empty space played an important part in their cosmogony, for the following reason. At the end of every cosmic cycle, when the hot element becomes predominant, the cosmos expands thermically and thus increases in volume. Now, the problem raised by this increase in volume can be solved, if the cosmos is, as assumed, an island in an infinitive void. The pupils of Aristotle, who, like their master, maintained that the cosmos was finite and denied the infinite extension of space, rejected this Stoic theory. The whole question became the subject of a scientific controversy between the two schools which bears a remarkable resemblance to the modern cosmological argument about space and the structure of the matter in it. We are told by a Stoic source: "Aristotle and his school maintained also that if there were a void outside the confines of the cosmos, all the matter would be poured out into the infinite and scattered and dissipated. But we maintain that this could never happen. For matter has a coherence which holds it together and against which the surrounding void is powerless. For the material world preserves itself by an immense force, alternately contracting and expanding into the void following its physical transmutations, at one time consumed by fire, at another beginning again the creation of the cosmos" [243]. The cohesion of the cosmos results from the tension of the pneuma, as we know from Stoic doctrine. This cohesive force, which binds the parts of the cosmos together into a single entity, offsets the dissipating influence of the surrounding infinite void and makes the cosmos a closed universe whose unity is not vitiated by changes in its size. There is an unmistakable analogy here with the theoretical argumentation of modern relativistic cosmologies. Once again we see that the inner logic of scientific patterns of thought has remained unchanged by the passage of centuries and the coming and going of civilizations: the same models and associations recur, only in new forms suited to the more advanced stage reached by physical knowledge.

IX

THE BEGINNINGS OF ASTROPHYSICS

"Where is the way to the dwelling of light, and as for darkness,
where is the place thereof."   JOB 38.19

A FTER the golden period of Greek science in the third and
second centuries B.C., a rapid decline of scientific creativeness
set in. Eventually, after some slight revivals of secondary importance, this decline merged into the general decay of ancient
culture from the fourth century A.D. onwards. From the time of
Aristarchus, Archimedes and Eratosthenes in the third century
B.C., and of Hipparchus in the second, no more important original
contributions were made in the two main branches of ancient
science, mathematics and astronomy. Ptolemy in A.D. 150 was
mainly a summarizer and interpreter of previous astronomical
doctrines, though he also worked at geometrical optics, as did his
contemporary Hero. Nor was there anything comparable to the
contribution of Epicurus and the Stoics to scientific thought after
Lucretius and Poseidonius in the first century B.C. The scientific
content of subsequent philosophical doctrines was weak and un-
original. On the other hand, the first century A.D. marks the
beginning of the work of compilers and interpreters which went
on for more than four hundred years and which is the mirror
wherein we see a large part of ancient Greek science. The
eclectic tendency which first appeared in philosophy in the first
century A.D. and led to the blurring and intermingling of the
doctrines of different schools also left its mark upon the scientific
viewpoints associated with these schools. No doubt the devoted

adherents of each school adopted its particular physical theories too; but in the general intellectual consciousness of educated circles at that time these theories began to be mixed up together, so that Stoic physics was combined with Aristotelian, and even pre-Socratic, elements. This was brought about by the tremendous achievements of "pure" science in the Alexandrian period. The astronomical measurements and calculations of Aristarchus, Hipparchus and Poseidonius, the determination of the earth's circumference by Erastosthenes, the expansion of geometrical knowledge—all this reinforced the tendency to be critical of the accepted view of the cosmos given by the philosophical schools, and encouraged the eclectic approach.

There is a scientific essay, written in about the year A.D. 100, which has preserved for us a vivid and colourful picture of the scientific opinions and arguments current at that time in intellectual circles in Athens and Rome. Here we find, as it were, a distillation of the science, philosophy and mythology of all the four hundred years that separate Empedocles from Poseidonius. The author is Plutarch, and the title of the essay (one of his philosophico-physical works) is "On the Face in the Moon". In form it is a conversation between several men who belong to different philosophical schools and who are all versed in the history of Greek science from the earliest times. In content it is what in modern terminology we should call an essay in astrophysics, perhaps the first work on astrophysics ever written. Astrophysics applies the methods and conclusions of physics to astronomy and is thus principally conditioned by the practical and theoretical attainments of physics. To-day thermodynamics, quantum theory, spectroscopy and nuclear physics all assist the scientist in comprehending the structure of the sun and the planets, the various classes of fixed stars and nebulae. Plutarch's essay deals with lunar physics. It employs arguments drawn mainly from dynamics and optics which are remarkable for their advanced grasp of gravitation and as marking the transition from geometrical optics to physical optics. There is a strange contrast between the last chapters of the book, which discuss the mythology of the moon and particularly its function as the repository of the souls of the dead, and the other parts which are distinguished by the clarity and acumen of their scientific reasoning in the best tradition of the time, as we find it, for example, earlier in the poem of Lucretius and later in the works of Ptolemy.

The scientific discussion in the book starts from the cosmological views of the Stoic Poseidonius (135-51 B.C.), which also contained a theory about the physical nature of the moon. Poseidonius, like Aristotle, distinguished two different zones in the cosmos—one between the earth and the moon and the other above the moon. The realm of air extends from the earth to the moon. The lower part of this, close to the earth, is composed of vapour and comprises the atmosphere, the height of which Poseidonius estimated at forty stadia, i.e. about seven kilometers. From there to the moon (Poseidonius was not far out in his estimation of its distance from the earth) there is an expanse of pure air. Beyond the moon begins the realm of fire in which revolve Venus, Mercury, the sun and other planets, and finally the fixed stars. The moon, revolving as it does on the boundary between the terrestrial realm of earth and air and the celestial realm of the stellar whorls of fire, is, according to Poseidonius, "a mixture of air and soft fire" [240]. Just as dark spots form on the sea when the wind ripples its surfaces, so the air makes parts of the moon's surface dark, thus giving it the appearance of a man's face. This view is opposed by Plutarch, who tries by all manner of arguments to prove that the moon is a body of the same kind as the earth. Since this opinion is at odds with the Aristotelian tradition which concentrates all the heavy bodies about the earth, the centre of the universe. Plutarch expounds something in the nature of a general theory of gravity. In this he is apparently not a little influenced by the Stoic doctrine of the pneuma, that pneuma which acts as a cohesive force upon all the parts of the cosmos everywhere. "If all heavy bodies converge to one and the same point, while each presses on its own centre with all its parts, it will not be so much *qua* centre of the universe as *qua* whole, that the earth will appropriate weights, because they are parts of itself; and the tendency of bodies will be a testimony, not to the earth of its being the centre of the universe, but, to things which have been thrown away from the earth and then come back to it, of their having a certain community and natural kinship with the earth. Thus the sun attracts all the parts of which it is composed, and thus the earth draws the stone to itself and makes it part of itself. . . . But, if any body has not been allotted to earth

from the beginning, and has not been rent from it, but somehow has a constitution and nature of its own, as they would
maintain to be the case with the moon, what is there to prevent
its existing separately and remaining self-contained, compacted
and fettered by its own parts? For not only is the earth not proved
to be the centre, but the way in which things here press and come
together towards the earth suggests the manner in which it is
probable that things have fallen on the moon, where she is, and
remain there" [260]. The inclusion of the sun, as well as the
moon, in this description of general gravity may lead us to assume
that the Aristotelian distinction between heavy and light is
completely annulled here. However, the next words of the
passage show—if they are not written simply to score a point—
that Plutarch's primary intention was to explode the conception
of a *single* centre and instead to postulate *many* centres of attraction, regardless of their "kind". "Why do not those scientists, who
compress all the heavy earthy bodies into one place and make
them parts of one body, impose the same law upon the light bodies
too? Why do they leave such enormous systems of fire separate?
Why do they not concentrate all the stars in one place by plainly
stating that all the fiery bodies that tend upwards must be part
of the same body?"

As has been noted, this transformation of the centre into a
general and relative conception was partly due to the pneuma
which permeates the cosmos, everywhere creating tensions and
cohesive forces. (We should bear in mind that Poseidonius gave
the idea of the pneuma fresh depth and vigour by his conception
of cosmic sympathy.) But the doubt which is specifically cast upon
the earth's being the centre of the universe points to a further
influence, namely the heliocentric theory of Aristarchus which
preceded the "Copernican revolution" by eighteen hundred
years, and incidentally was also conjectured by others in the
Ancient World. This theory did not convince the contemporaries
of Aristarchus (middle of the third century B.C.) and was finally
rejected by no less an authority than Hipparchus (middle of the
second century), who took his stand upon the accepted geocentric
view. It is, therefore, of interest to find in the above quotations
proof that the heliocentric idea was not forgotten, but still exerted
an influence as late as Plutarch's time and was instrumental in

shaking the accepted belief that the universe had an absolute centre. Elsewhere in the debate Aristarchus is mentioned by name: "Do not bring against me a charge of impiety such as Cleanthes used to say that it behoved the Greeks to bring against Aristarchus of Samos for moving the Hearth of the Universe, because he tried to save the phenomena by the assumption that the heaven is at rest, but that the earth revolves in an oblique orbit, while also rotating about its own axis" [235]. It should be remembered that these words were written only a few decades before Ptolemy's *Syntaxis*, the book which established the geocentric theory for fourteen hundred years and, also in accordance with ancient tradition, maintained the finiteness of the cosmos. Aristotle had reached the conclusion that the cosmos is finite from the premiss that the earth rests at its centre, since there can be no centre to an infinite expanse. In Plutarch we find this line of argument reversed. The first influence at work here is the turning of the centre into a relative concept; the second is the cosmology of the Stoics, who, while maintaining the finiteness of the cosmos, nevertheless postulated the existence of the infinite void stretching beyond it. "How and of what can it be said that the earth is at the centre? For the universe is infinite; and the infinite, as having no beginning and no limit, cannot be given a centre. For the very idea of centre implies a limit, whereas infinity is the negation of every limit" [261]. This was the view, in the fifteenth century, of Nicholas of Cusa, who believed that the universe was infinite. This whole line of argument and its converse are most instructive in demonstrating how the idea of the absolute centre and the finiteness of the universe are interdependent: overthrow one, no matter which, and you inevitably overthrow the other.

The Aristotelian assumptions that heavy bodies tend to the centre and that the earth is round together gave rise to questions which at that time seemed paradoxical. Pliny (A.D. 23-79) in his *Natural History* summarizes some findings which had been arrived at before his day: the fact that the surface of the ocean is part of the surface of a globe, seeing that the surface of the water is vertical to the direction of gravity; and the notion of the antipodes which, in relation to us, are upside down. He also points out that the height of mountains (he specifically mentions the Alps) is of no consequence in comparison with the radius of the earth, i.e. that the unevennesses of the earth's surface do not in fact vitiate its sphericity. These conclusions are known to the disputants in *The Face in the Moon.* They themselves add several others concerning the internal force of gravity *within* the earth's orb. Some of these call to mind classic exercises from Newton's Theory of Gravitation. "We must not listen to philosophers, if they claim to meet paradoxes with paradoxes, and controvert surprising doctrines by inventing others still more strange and surprising, as these people do with their idea of motion towards the centre. What absurdity is there that this does not imply? Does it not mean that the earth is a sphere, though it contains such enormous depths and heights and irregularities? That people dwell at our antipodes, like wood-worms or lizards, clinging to the earth with their lower limbs upwards? That we ourselves do not remain perpendicular as we walk, but remain at an angle and sway like drunken men? That masses weighing a thousand talents, borne down through the depth of the earth, come to rest when they reach the centre, though nothing meets or resists them; and that, if in their downward rush they should overshoot the centre, they would turn back again and reverse their course of themselves; . . . that a rushing stream of water falling downwards, if it came to the centre, which they themselves declare to be incorporeal, would halt suspended round it, or circle about it, oscillating with an oscillation which never stops and never can be stopped? Some of these things a man might, without perjuring himself, force himself to represent as imaginable by his intelligence. But this is making up down, and everything topsy-turvy, with a vengeance, if things from us to the centre are 'down', and things under the centre are 'up' again. This would mean, for instance, that, if a man through sympathy with the earth should stand with the centre of the earth at his middle, he would have his head up and his feet up too" [259]. Here we have scientific doctrine enunciated in the form of paradoxes supposedly resulting from a certain scientific theory. But the form is not of the slightest importance. What does matter is the transformation observable here of the Aristotelian idea of the downward tendency of heavy bodies into a chapter in the dynamics of the earth's gravitational field. This process, as already noted, is a consequence of two things: the views of the Stoics, especially Poseidonius, and the scientific achievements of the third and second centuries. There is no doubt that the findings of the mathematicians and astronomers during these centuries had an appreciable effect upon the intellectual's picture of the cosmos at the end of the classical period. The introduction of geometrical dimensions into space, the numerical determination of the distances and the sizes of the chief celestial bodies, the first conception of "astronomical distances" and their comparison with terrestrial measurements— all these gave rise to a widespread tendency to examine the heavenly phenomena from the standpoint of reason. It might be called a revival of the approach of Anaxagoras, who had surmised that the sun and moon were merely fiery stones and conjectured their distance from the earth. Only by Plutarch's day this rationalizing process had a foundation of scientific fact which did not yet exist in the fifth century B.C. It is likely enough that this scientific progress aroused the same feeling of distance and objectivity with regard to celestial phenomena as marks the beginning of the modern period.

Plutarch makes use of astronomical arguments in the ninth chapter of *The Face in the Moon*, where he cites figures from Aristarchus' book on sizes and distances. Aristarchus proved that the distance of the sun from the earth is more than eighteen times, and less than twenty times, the distance of the moon from the earth (this proportion is actually twenty times too small). From this Plutarch argues that the moon certainly "belongs" to the earth and that the closeness in space reflected by those relative astronomical distances proves also a "family" closeness, i.e. the earthly nature of the moon. He uses similar arguments in another place (chapter 6) against those who question the "heavy" character of the moon. We know, he says, from the shadow thrown by the earth across the moon during an eclipse that the earth is much larger than the moon, and yet it hangs poised in the void without any support; obviously then, *a fortiori*, this can be true of the smaller moon. But he is not satisfied with these arguments. He has to set at rest any fear of a heavy body's falling upon the earth's surface. This he does by arguments from physics. "The moon has a security against falling in her very motion and the swing of her revolution, just as objects put in slings are prevented from falling by the circular whirl; for everything is carried along by the motion natural to it if it is not deflected by anything else. Thus the moon is not carried down by her weight, because her natural tendency is frustrated by the revolution. Nay, there would, perhaps, be more cause for wonder if she were altogether at rest and unmoved like the earth" [258]. Here we have another passage that anticipates the seventeenth century. The movement of the moon is the sum of two components; if it stopped revolving, it would begin to fall by gravity towards the earth. The choice of the stone in the sling as the analogy of the moon's movement shows us the extent to which there was a consciousness of the unity of physical phenomena. The sentence about the continuation of a body's movement as long as it is not deflected is in fact the formulation of the law of inertia given by Galileo and Newton more than fifteen hundred years later.

Of course, we must beware of hasty generalizations when we come across a passage such as that just quoted. Intuitions of this kind were few and far between in the Ancient World; nor could they, in the absence of systematic research on a sound theoretical and practical basis, be combined into a single homogeneous system. Still, the general tendency is plain enough; Aristotle's schematic categories are beginning to disintegrate as the approach to physical phenomena becomes more flexible and less bound by accepted notions. The bonds have not yet been completely severed, however, as we see from the attempt to explain the conjecture about the moon's earthly nature within the framework of the Aristotelian terminology of natural place, "up" and "down". The moon's movement in the sky, so far from proving that it is different in nature from the earth, points to the conclusion that there are bodies of the same kind as the earth which are not in their natural place. So Plutarch argues, and supports his view by a parallel taken from another element, fire. The fire shut up in the bowels of the earth under the volcanoes is still fire, even though it is out of its natural place. Again, the air confined in a bellows is a "light" element with an upward tendency, but for all that it is forced to flow elsewhere than to its natural place. To make his case still stronger, Plutarch goes all the way back to the great authority of Empedocles. This appeal is not only typical of the eclectic spirit of Plutarch's time, but demonstrates the closeness of its intellectual viewpoint to the scientific approach of the great pre-Socratic philosophers. Plutarch maintains that we must guard against too literal an interpretation of the expression "natural place". "Take care, my friend, not to remove everything to its natural place, for this is to insist on the disintegration of the cosmos by introducing the Empedoclean 'conflict' into matter" [262]. This absolute separation of different categories into insulated compartments would, in Plato's words, create a world without God, something like a body devoid of mind and soul. The world cannot exist without that antithesis which Empedocles called Love "that brings about change of place and the transference of qualities from body to body, that imposes the necessity of movement on one and the necessity of rest on another, that forces a body to move from its natural place to a better one thus creating the harmony of the universe". This criticism of natural place logically leads to a critical examination of the concept of "natural motion". It involves a negation of a fundamental distinction of Aristotelian physics. In order to explain the eternal circular motion of the heavenly bodies in opposition to the movements of heavy and light bodies, Aristotle postulated that they were composed of a fifth element, the aether. Plutarch reverts to the terminology of Anaxagoras which identifies fire and aether. After discarding Aristotle's main idea of assigning a special status to the stars, Plutarch questions the validity of natural motion: "If all bodies were bound by their natural movements, the sun would not move in a circle, nor would Venus nor any of the other stars. For, by their nature, light and fiery bodies should move upwards and not in a circle. But if nature permits such differences in change of place (that on earth fire is borne upwards, whereas when it reaches the sky it revolves in a circle), it is not surprising that heavy earthy bodies too, on reaching the sky, should be compelled to change their form of motion" [263].

Plutarch eventually reaches a conclusion which renders the whole of Aristotelian dynamics null and void. "If we give up the theories to which we have become slaves of habit and fearlessly state what we think, then it would appear that no single part of the universe has its own peculiar order or position or movement which can be categorically called natural. Instead, we must say that if everything moves in a way appropriate to that through which it exists and for which it was created, and if it is set in motion and functions and is affected in accordance with the needs

212

of that other's preservation or beauty or power, then this is what is meant by a natural place, a natural motion and a natural condition" [264]. In other words, no such antithesis as natural and unnatural motion exists, since the motion of every body is natural in the context of the special circumstances which gave rise to it or which placed the body in its given condition and place. With these words Plutarch destroyed the foundation of Aristotelian physics. But he did not erect in its place any other system of physics which would make it possible to analyse those special circumstances and show how the various forms of motion result from them and "conform to them". It was not enough in a flash of intuitive genius to hit upon several details of the Newtonian law of gravity or hint at the law of inertia, as in the isolated passages quoted above. What was needed at this point was the construction of a new analytical system which should lead criticism out of the cul-de-sac of generalities on to the high-road of experimental science. Instead of this, we find allegory: Plutarch, in the spirit of Plato, compares the cosmos to a living creature. Just as a man's limbs are not arranged according to natural principles in Aristotle's sense, so the parts of the cosmos are arranged as befits the laws and requirements of a living organism. In chapter 15 each different heavenly body is compared to a limb of the human body: the sun to the heart of the universe (after Poseidonius), pumping out heat and light like blood; the earth to the bowels; the moon to the liver, which lies between the heart and the bowels. At the end of the chapter it is once more stressed that the bodies are not balanced through their weight but on another principle of arrangement. Here the Stoic idea that the universe is a single, homogeneous organism is carried to the extreme of a detailed analogy. At the same time, the Platonist in Plutarch rejects the Aristotelian rule, accepted by the Stoics, that the heavy elements tend downwards while the light rise upwards. More important, however, than all these details is the actual comparison of the cosmos to a living body. Here we may find a clear indication of the Weltanschauung of the Greeks. The point is so significant that we shall have to return to it later.

We shall now consider those chapters of *The Face in the Moon* which deal with the nature of the moon as demonstrated optically. The foundations of geometrical optics had already been discussed by Euclid in his work on Light (*c.* 300 B.C.). Hero of Alexandria published his book on Mirrors shortly after Plutarch's time, but it is likely that many of these problems were already known several centuries before Hero, in the time of Archimedes and perhaps even earlier. Hero also considers the question of the physical conditions of reflecting surfaces: in stressing the necessity of polishing the surface to obtain the best conditions of reflection, he attributes faulty reflection to the absorption of light rays by the pitting of an unpolished surface. Thus the laws of reflection, including the outcome of a repeated series of reflections from many surfaces (reflections from two or more mirrors), were known to Plutarch's contemporaries. They no doubt also understood the laws of refraction which were subsequently formulated in detail by Ptolemy in the second century A.D. A large part of Ptolemy's book has been preserved in a Latin translation made in the twelfth century from an Arabic version of the Greek original. Not only does Ptolemy undertake a detailed examination of the phenomena of refraction from a geometrical standpoint: he further explains the refraction of light in the atmosphere by changes in the density of the air. It is against the background of this contemporary knowledge of optics that we should judge Plutarch's use of optical evidence in support of his contention that the moon is earthly, as against Poseidonius' view that it is a mixture of fire and air. "There is no reflection from a rarefied or fine body. Similarly, it can hardly be supposed that light will be reflected by light, or fire by fire, for the reflecting body must be massive and dense, so that rays may strike it and be reflected from it. For example, the air transmits the very rays of the sun without checking or resisting them, whereas trees, stones and woven fabrics, when placed in the path of light, cause it to be reflected and scattered to a great extent. We see that the earth too is lighted by the sun. For it does not let the rays of the sun pass into its depths, as in water, or throughout its whole substance, as in air. But just as a circle crosses the moon and cuts off part of it, so is the surface of the earth cut into two in such a way that one half is lighted while the other remains unlighted. . . . Permit me to speak geometrically and in terms of proportion: if we see that, of the three things (earth, moon and air) which the sun's light strikes, the moon is illuminated not in the same way as the air but in the same way as the earth, it necessarily follows that these two objects which are affected by the same thing in the same way must be of a similar kind." [267]. Plutarch's view is further confirmed by the fact that during a solar eclipse the sun's light does not penetrate the moon. Poseidonius, it is true, tried to harmonize this with his own theory, but Plutarch rejects his argument: "The statement of Poseidonius that it is because of the depth of the moon that the light does not come through her to us is obviously wrong. For the air which is unlimited, and has a depth many times that of the moon, is lit up throughout by the sun and by its rays shining upon it. We are therefore left with the view of Empedocles that the illumination which comes to us from her is caused by a sort of reflection of the sun upon the moon. Hence it is that no heat or brilliance reaches us, as should be expected if there had been a kindling and mixture of lights. But, just as the echo sent back when voices are reflected is weaker than the original sound, and the blows struck by ricochetting missiles fall with less force, 'so the beam, striking on the broad circle of the moon', sends to us a feeble and faint stream of light because its force is dissipated owing to the reflection" [265].

The next question debated is the way in which light is reflected from the moon. These interesting chapters show us how the first concepts of physical optics began to develop at that time out of the geometrical theory of light. The opponents of the reflection theory of the moon's light start from the premiss that the only kind of reflection is the geometrical reflection of mirrors. This leads them to a twofold argument: ". . . According to the law that the angle of incidence is equal to the angle of reflection, it follows that, when the moon in the middle of the sky is half illuminated, its light will not be carried to the earth but will be deflected away from it. For the sun will then be on the horizon and its rays will strike the moon and be reflected from it at the same angle. In that case they will fall beyond the other extremity of the earth so that their light will not reach us. Otherwise there would have to be a large deflection of the angle (of reflection), which is impossible. . . ." The second argument of the opposition is as follows: "We take our stand on the assumption that whoever stands in the path of the reflected rays will see not only the illuminated body, but also the illuminating body. . . . Since this has been proved in theory and in practice, those who maintain that the earth is illuminated by reflected light from the moon must be able to show us at night the sun's image in the moon, just as we see it by day in water which reflects the sun's light. Since this image cannot be seen, they should conclude that the moon's light is not produced by reflection. Hence the moon is not of the same nature as the earth. . . ." These points are answered by the postulation of a *scattered* reflection in both solids and liquids. This can be explained by geometrical reflection, if we suppose that the reflecting surface is composed of many small mirrors in all manner of conditions. ". . . Many parts of the moon are not smooth but rough. This brings it about that the rays which reach us from such a large body and such a great height have been reflected criss-cross from all sides, blending and merging into a single brightness, just as if they reached us from many mirrors. . . . Milk too does not form an image like a mirror, nor does it reflect the likeness, since its parts are rough and not smooth. Why then should the moon be capable of reflecting an image like polished mirrors? Moreover, if there is a scratch or dirt or rough- ness on such mirrors at the point of reflection, the image, though still visible, will be blurred." [266, 269].

The discussion about the nature of the spots on the moon keeps recurring throughout the book. Can this be an optical illusion caused by the faintness of the moon's light? This contention is refuted by transferring the question from the observed to the observer. It is just short-sighted people who cannot distinguish any differences of shape on the moon's orb and see it as absolutely smooth. The unevenness of the moon's surface obviously rules out the theory of an optical illusion, since in that case the whole surface ought to be blurred. The Stoic conjecture—that the spots are air pockets within the moon's fire, or a layer of air on that fire—is contradicted by the permanent location of the spots in the same place. If masses of air were spread over the surface of the moon, they would cease to be visible to us when the moon is at its full and the sun's light strikes it most powerfully. Here Plutarch draws a parallel from the earth. Just as the sun's light penetrates into and illuminates all the air on the earth, except for that in valleys and recesses which remains opaque and un- illuminated, so we may suppose that the opaque spots on the moon are valleys which the sun's light does not reach. "And, as to this apparent face in her, let us suppose that, just as our earth has certain great depressions, so she is opened up by great depths and clefts containing water or dark air, which the light of the sun does not penetrate or touch, but is there eclipsed, so that the reflection sent hither is scattered" [268]. Finally, the size of the spots is estimated on the basis of what was known of the moon's dimensions; and the argument that such enormous shadows indicate that the objects which throw the shadows are equally large is refuted by showing that the length of the shadow depends also on the sun's position.

After the scientific debate and before the chapters on mythology which end the book, the disputants raise the further question whether the moon is inhabited either by animals or human beings. The very fact that such a question is raised and the manner in which it is discussed, give us further insight into the transformation which occurred in Plutarch's day of Aristotle's philosophical axioms into relative concepts. The question is raised in connection with the central theme of the discussion. Someone maintains that, if the moon were uninhabited, the case of those who deny its earthly nature would be strengthened, especially if the question is put in a slightly different way—Is the moon suitable for habitation by living creatures? May we, on the assumption that it is empty and desolate, argue that it was created in vain and for no purpose? Here we find an instructive parallel with a book written in modern times, Kant's *Naturgeschichte und Theorie des Himmels* (1755). The third chapter of this book, entitled "On the Inhabitants of the Stars", follows a similar line of reasoning to that contained in chapters 24 and 25 of *The Face in the Moon*. Both Plutarch and Kant cite the desolate areas of the earth in refutation of the argument that an uninhabited star has no value. "After all," says Plutarch, "we see that not all of the earth is fertile and inhabited. On the contrary, only a small part of it, such as the peaks and the peninsulas which emerge from the sea, produce living creatures and plants. The other parts are desolate and barren because of winter storms and erosion, or they are sunk under the ocean, as is the case with most of the earth" [270]. Plutarch goes on to make a comparison in the spirit of the Stoic doctrine of cosmic sympathy. Just as the uninhabited regions of the earth have a beneficial influence upon the inhabited parts (vapour from the sea, the melting of the snows in northern climes, etc.), so it may be that the moon, even if uninhabited, has an influence upon us by the radiation of the light reflected from it (which is made up of the rays of all the other stars as well as the sun's). Plutarch here is influenced by Poseidonius' feeling for the cosmic. Kant, on the other hand, with his awareness of the infinite cosmos of Newton, points out how much more insignificant one planet is in the whole of the creation than a desert or desert island in relation to the earth's surface. Plutarch does not rule out the possibility of the moon's being inhabited, just as Kant presumes that there is life on the planets. But both of them come to the conclusion that such creatures will have a different physical structure from man. Plutarch cites the multiplicity of forms adopted by creatures on our earth and their adaptation to different conditions: in the same way it may be supposed that the creatures on the moon are adapted to the conditions of life there. At the end of the discussion he imagines the inhabitants of the moon looking at the earth and asking themselves how this dark, damp mass, full of cloud and mist, can produce and support life, and concluding, perhaps, that the moon is the only earth.

To complete the picture of "the scientific climate" of Hellenistic culture at the beginning of the present era, we shall briefly mention another astrophysical subject which is discussed in the *Quaestiones Naturales* of Seneca (*c.* 3 B.C.-A.D. 65). This is the nature of comets. The shape of the comets and the manner of their appearance has in all generations been a source of baffled wonderment. Aristotle in his *Meteorologica* regards them as meteorological phenomena which form in the sublunary region between the earth and the moon. Within the framework of his own understanding of the cosmos, his arguments are reasonable enough: the comets lack that permanence and constancy which are the outstanding signs of the eternity and divinity of the other stars; they suddenly appear in the sky, sometimes moving at great speed, and disappear after a short time; their course is not within the signs of the zodiac, unlike all the other planets. Aristotle's view was accepted by the astronomers, who went on regarding the comets as atmospheric phenomena up to the end of the sixteenth century. It was Tycho Brahe who, in an essay written in 1588, first proved by his observations and measurements that the great comet of 1577 was beyond the orbit of Venus. For all that, Galileo still upheld Aristotle's view, which was finally discarded only in Newton's generation. Seneca presents the opinions of two scientists, Epigenes and Apollonius of Myndos, the former of whom was a follower of Aristotle, while the latter (apparently a contemporary of Seneca's) opposed him with a theory which is close to the modern view. Seneca is inclined to support Apollonius, as is clear from his introduction: "To clarify this problem we should first enquire whether the comets are similar in kind to the upper stars. Now, it is true that certain features are apparently common to both: their risings and settings, and their shape too, for though that of the comets is looser and longer, still both consist of bright masses of fire. . . . For this purpose, it is essential to collect information about comets from the earliest times. For till now it has been impossible to ascertain their courses because of their rarity, or to discover whether they keep to a fixed cycle and whether their appearance on a given day is the result of a fixed sequence." Seneca goes on to describe Epigenes' theory that comets are formed as a kind of fire swept on high in a whirlwind. Seneca objects that, if this were so, comets would only appear during storms, whereas in fact they are also seen when there is no wind at all and the fluctuations in their brightness are in no way related to changes in the strength of the wind. Winds strike wide expanses of air, whereas a comet appears only in one place and at a height beyond the reach of winds. Finally, there is no resemblance in shape between a whirlwind and a comet.

Next, Seneca quotes the conjecture of Apollonius of Myndos: ". . . A comet is a separate star like the sun and the moon. In shape it is not compressed into a ball, but loose and elongated; it has no definite course, but it passes across the upper regions of the cosmos and becomes visible only on reaching the lower part of its course. . . . The comets are many and various, and different in size and colour: some of them are red and dull; the light of some is bright, clear and liquid; the flame of others is not clear and soft, but enveloped in lurid smoke. Some of them are blood-soaked and terrifying, ominous of future bloodshed. Their light waxes and wanes, like other stars that are bright when setting and large when close to us, but grow small when they rise higher and shine more dully when they are further away" [256]. Seneca agrees in principle with this view. He dismisses the view of his contemporaries and fellow Stoics that a comet is a sudden flame. He is sure that it is one of the eternal creations of nature, as against the transient phenomena of the atmosphere. "How can anything persist in the air, if the air itself does not persist in one state for long?" Fire tends upwards, whereas the course of the comet is curved, like that of the other stars. "I do not know whether other comets follow such a course, but two of them in our time have done so." At the same time Seneca has doubts which qualify his acceptance of Apollonius' view on certain points. He mentions the old argument that comets are not confined within the region of the zodiac. Then he has another point: "A star can never be seen within another star. Our eye cannot penetrate the star and see the upper regions through it. But through a comet it is possible to discern the portions of the sky above, just as through a cloud. From this it would follow that a comet is not a star, but light and unstable fire." It is clear to Seneca that the time has not yet come to pass final judgment on the nature of comets. "Why should it surprise us that so rare a cosmic phenomenon as the comets cannot be brought within the framework of regular laws, and that we cannot know their beginning or their end, since they reappear only after such enormous periods of time? . . . The day will come when time and the researches of long generations will bring to light what is now concealed. A single generation is not enough for the solution of such great problems. . . . Then someone will arise to explain the courses of the comets, why their paths are different from those of the planets, and what is their size and nature. Let us be content with what we have discovered so far. Those who come after us will also add their share to truth" [257]. It is hard to imagine a greater contrast to Aristotle's apodictic statements than these noble words of Seneca's. They are perhaps the last among the few sparks of awareness of scientific progress which we meet with in antiquity. Two earlier ones with a time interval of about two hundred years between each of them might be mentioned here.

The first occurs in the second chapter of *Ancient Medicine*, a medical treatise dating back as far as 420 B.C. It reads as follows: "Full discovery will be made, if the enquirer be competent, conduct his researches with knowledge of the discoveries already made, and make them his starting-point." The other passage is to be found in Archimedes' *Method*. Before stating his theorem on the volumes of a prism and a pyramid, Archimedes says: "I am persuaded that it will be of no little service to mathematics; for I apprehend that some, either of my contemporaries or of my successors, will, by means of the method when once established, be able to discover other theorems in addition, which have not yet occurred to me."

Although the history of the various sciences had been developed as a subject for study as early as the time of Aristotle and his disciples, it was no doubt only the great scientific period of the third and second centuries B.C. which brought about the first beginnings of a more permanent and more widespread cognizance of scientific progress and its significance. Seneca's words faithfully reflect the feeling of his generation, which, like us to-day, saw the understanding of the cosmos as an historical process, an endless task passed on from generation to generation.

X

LIMITS OF GREEK SCIENCE

"Hast thou perceived the breadth of the earth? Declare if thou knowest it all." JOB 38.18

THE picture unfolded by a general survey of Ancient Greek Science is characteristic of the birth and death of a living organism. We witness the period of germination in the sixth and fifth centuries with the appearance of the Milesian School, the work of Pythagoras and his first pupils, and the teachings of Empedocles and Anaxagoras. This seed comes to fruition in the period from Leucippus and Democritus and the later Pythagorean School in the second half of the fifth century to the death of Archimedes at the end of the third century. In the second century, after Hipparchus, the pace of creation becomes noticeably slower and the long-drawn decline of creative power begins. Its place is taken by the activity of compilers and commentators, beginning in the first centuries of the Christian era and continuing until the final eclipse of classical culture. The whole process covers a long enough period: about eight hundred years if reckoned from Thales to Ptolemy, and more than a thousand if we carry it down to the time of the later commentators. In this present review of the Greek cosmos the world of organic life has not been included. But even if biology were included in it, there would still be no doubt that the main contribution of the Greeks to the human heritage was made in the sphere of astronomy and mathematics, especially geometry. In the other physical sciences there are no achievements proportionate to the vigour of the first

222

outburst of scientific activity and the length of time which that activity lasted. Very few indeed are the quantitative laws that were formulated in all the branches of "terrestrial" physics: if we mention Pythagoras' laws of musical harmony, Archimedes' law of leverage, and some of Hero's laws of geometrical optics, that is practically all. Similarly very little progress was made in dynamics. While not wishing to belittle Aristotle's achievements in this sphere, we must remember that his whole teaching is only partially quantitative; and after him the only name is that of Hero, who formulated the laws of the parallelogram of velocities. The important work of Archimedes on hydrostatics and the centre of gravity is of a static nature, as is his law of leverage. Side by side with the absence of quantitative formulation, we are struck by the lack of accurate measuring instruments and the slow development of simple machines. This brings us to one of the strangest and most puzzling phenomena of history—the absence of technology in Ancient Greece. We have grown so accustomed in the modern world to regarding science and technical progress as inseparable that we cannot understand how the nation which, by discovering the scientific method, paved the way for modern science failed to display any initiative whatsoever in the technical sphere. It can be said that the slow progress made by Greek science, apart from astronomy, completely belies the greatness of its vision and its original momentum, and that its few technical contributions fall far short of its scientific achievements.

We cannot discuss these questions without touching upon sociology and psychology and eventually entering into historical considerations which belong rather to the broad history of culture that to the narrower history of science. Although, in recent years, a number of valuable works of research have helped to clarify the question, we can hardly regard it as solved or provide a completely satisfactory answer to it. In this investigation it is helpful to compare the Greek period with our own, even in simple matters. We shall begin with these. One of the factors which retarded the progress of Greek science was undoubtedly the isolation of the man of science which from time to time snapped the chain of evolution. Such isolation would also seem to have been a fact at the beginning of the modern period. Only a very small handful of men showed intelligent interest in Copernicus' work and gave him any encouragement. Much the same was true of Tycho Brahe and Kepler. But there was a fundamental difference. The modern scientists were favoured by the existence of universities and the general atmosphere of learning in Europe. These, together with the spread of printing, led to the beginnings of the organization of science in the middle of the seventeenth century, with the founding of the first academies in Italy, France and England. It may be asked if Plato's Academy and the other philosophical schools did not perform a similar function in antiquity. While it is true that in these institutions attention was also paid to the natural sciences and mathematics, these subjects were studied as part of the specific doctrine of each philosophical school and were of secondary importance to the actual philosopical teaching. Thus there was no uniform atmosphere in which a continuous tradition of constant scientific progress could flourish. This brings us to another point that must be taken into account. Greek science and modern science alike have their origin in a revolutionary departure from what preceded them. The Milesian School opposed logos to mythos, while Galileo and the investigators of the seventeenth century set science free from the swaddling-bands of the Church and made it an independent sphere of human thought. The fundamental difference between these two historical processes is that, whereas the former *tied* science to philosophy, the latter *untied* the bonds binding them together. When modern science turned its back on scholastic philosophy, and the philosophy of Aristotle,. it simultaneously turned its back on all philosophy. Galileo and his pupils, the first members of the Royal Society in London, Newton, and Huyghens in Holland—all the founders of seventeenth-century science— were investigators of nature, not philosophers. After Descartes and Leibniz there were no philosophers who contributed anything of importance to the exact sciences. That was the parting of the ways. To-day the occasional contact between philosophers and scientists takes the form of an epistemological discussion confined to the meaning of the attainments of science and does not affect its methods.

In antiquity, on the other hand, it was Greek philosophy that brought about the transition from mythology to rational thought. It is true that at first the central problem of this philosophy was to explain natural phenomena by rational causes. But this aim quickly merged with the wider impulse to investigate systematically the various forms of knowledge, the fundamental nature of reality and the place of man in the world. This is the way in which the philosophical schools became centres of scientific research. Only after Aristotle, in the Hellenistic period, did there arise professional scientists in the modern sense, great investigators, mathematicians and astronomers, such as Euclid and Archimedes, Aristarchus and Apollonius of Perga, Eratosthenes and Hipparchus. Even so, Greek science was still overshadowed by philosophy, and for two reasons. First, because of the tremendous educational influence of Plato, who was the inspiration of astronomical and mathematical research; and secondly, as a result of Aristotle's encyclopedic systematization and the great influence of the Stoic School in the centuries after him. Since the central question in philosophy is "Why?", and "How?" is subordinate to it, this long association of philosophy with natural sciences was harmful to the progress of the latter. Logic and deduction were more important than induction and experience, and the teleological view of nature hampered the increase of physical knowledge. The history of modern science has taught us how important for the understanding of nature can be, the proper selection of some sometimes apparently insignificant phenomenon, out of the enormous mass of phenomena, and the unflagging observation of its smallest details. Two famous examples of this from the seventeenth century are Galileo's discovery of the laws governing falling bodies, and Newton's explanation of the spectral decomposition of light. The philosophical approach is unfavourable to such reasoning from the particular to the general: the fundamental problems of philosophy are essentially general and it builds up its system by integration rather than by differentiation. Hence the paradox: philosophy performed the historical task of bringing a scientific attitude into the study of nature, but at the same time it was one of the factors that hindered the most advantageous development of this study.

Now we must consider the more ponderable influences. The problem of Greek technical achievements—or rather the absence of such achievements—has received much attention and all sorts of explanations have been proposed. In comparison with the peoples of the East, especially Egypt, we cannot find any noticeable technological progress during the Greek period. The

Egyptians, in the course of thousands of years, developed systems of building in stone on a scale never since equalled; at the same time they perfected methods of resolving all the incidental problems involved in this work, such as the quarrying, transport and raising of enormous stones, the erection of pillars and gigantic obelisks, the details of which have exercised the minds of engineers and historians in recent centuries. These marvellous methods completely overshadow the achievements of the Greeks in building, and even those of the later Romans. The requirements of building led to the invention of machines based on the fundamental principles of mechanics—the lever, the inclined plane, and other devices for reducing the effort at the expense of the distance traversed. Among all the other achievements of the Egyptians, we may mention their technique of mining, which enabled them to exploit production to a great depth, and their new methods in metallurgy. Slavery did not hamper these technical developments: on the contrary, the question of how best to use great masses of men in large technical undertakings raised further technical and organizational problems which were most successfully solved by the Egyptians, as we see both from the results and from the various descriptions given by the organizers themselves. This technical progress of the Egyptians refutes the view that the existence of slavery in Ancient Greece was the decisive reason for the lack of technological development. The opponents of this view have rightly emphasized that it is based on an exaggerated estimation of the economic scope of slave-owning. At the same time, they do concede that a psychological factor was here at work: contempt for slaves also involved contempt for what slaves do, namely, manual work. Basically, Greek mentality was *aristocratic*. Through judging the value of manual labour by the social status of the slave who performed it, the Greek came to despise it as unworthy of man's spiritual destiny. Plato says: "Why, again, is mechanical toil discredited as debasing? Is it not simply when the highest thing in a man's nature is naturally so weak that it cannot control the animal parts" [129]. And Aristotle says in his *Metaphysics*: "Hence we think also that the master-workers in each craft are more honourable and know in a truer sense and are wiser than the manual workers, because they know the causes of the things that are done (we think the manual workers are like certain lifeless things which act indeed, but act without knowing what they do, as fire burns—but while the lifeless things perform each of their functions by a natural tendency, the labourers perform them through habit)" [167]. Habit is the uncomprehending acquisition of experience. Therefore the artist or manager who is capable of imparting his knowledge to others is superior to the workers. Further on in the same passage we can see the connection between the contempt of the Greek for manual work and his rejection of the practical application of science. The admiration shown to outstanding artists, Aristotle maintains, was not due to their inventing something useful but to their wisdom and unique talent. "But as more arts were invented, and some were directed to the necessities of life, others to recreation, the inventors of the latter were naturally always regarded as wiser than the inventors of the former, because their branches of knowledge did not aim at utility" [167]. Aristotle's philosophy of history also displays this aristocratic mentality: "Hence when all such inventions were already established, the sciences which do not aim at giving pleasure or at the necessities of life were discovered, and first in the places where men first began to have leisure. This is why the mathematical arts were founded in Egypt; for there the priestly casts was allowed to be at leisure" [167]. The process of thought is clear enough. Work without study is worthless, whereas study for its own sake is the highest level of spiritual activity, superior to the combination of study with any practical purpose. The value of science is reduced when it becomes a means to an end. Most instructive is the comparison that Aristotle draws between the various kinds of science and a man's social status: "But as the man is free, we say, who exists for his own sake and not for another's, so we pursue this as the only free science, for it alone exists for its own sake" [168].

In addition to all the explanations already given for this preference of theoretical science to practical, i.e. experimental and technical science, there is a further important reason which derives from the character of the Ancient Greek: he saw no need to improve any further upon the technical achievements known to him. Aristotle once again reveals this psychology to us in his philosophy of history, as expressed in the passage already quoted from the first book of the *Metaphysics*. "That it is not a science of production is clear even from the history of the earliest philo-sophers. For it is owing to their wonder that men both now begin and at first began to philosophize; they wondered originally at the obvious difficulties, then advanced little by little and stated difficulties about the greater matters, e.g. about the phenomena of the moon and those of the sun and of the stars, and about the genesis of the universe. And a man who is puzzled and wonders thinks himself ignorant . . . therefore since they philo-sophized in order to escape from ignorance, evidently they were pursuing science in order to know, and not for any utilitarian end. And this is confirmed by the facts; for it was when almost all the necessities of life and the things that make for comfort and re-creation had been secured, that such knowledge began to be sought" [168]. Compressed into these few sentences is the whole mentality of the Ancient Greek with regard to the relative values of the basic goods of life. First of all comes study for its own sake, investigation for the sake of knowledge, not for the improvement of the conditions of life. Indeed, in Aristotle's view, nothing much more can be done in this direction: technical progress has already reached the level at which it can supply the essential necessities of life, and one of its most important con-sequences is pure science and philosophy. Technical improve-ments, if they are necessary at all, have no value in comparison with man's capacity for wonder which impels him to master the secret of the cosmos. Man's most priceless possession is his pure intellectual curiosity. Hence, the value of a discovery is in no way enhanced by its practical and technical possibilities. The world as created and the place that man has found for him-self in it provide him with all that he requires materially for his spiritual life and the maintenance of his transcendental values.

One more basic Greek attitude to life gave added impetus to this trend of thought: the insistence on moderation, especially in material requirements. Everything beyond a "bare necessity" was considered a luxury which was not proportionate to the work lavished upon it. It is a mistake to think that this attitude was a monopoly of Platonic philosophy which was taken over by Aristotle. Even a philosopher like Democritus, whose whole view of the cosmos is as different as it could be from Aristotle's, stressed the necessity of restricting man's inclination to devote too much of his limited strength to improving his material conditions. "One should realize that human life is weak and brief and mixed with many cares and difficulties, in order that one may care only for moderate possessions, and that hardship may be measured by the standard of one's needs" [91].

As early as the mythological period, a deep imprint had been left on the mind of the Greek by his awareness of the limits set upon man's power against forces beyond his control, and his fear of the vengeance that would be taken by these forces, in the event of his transgressing these limits. Such presumptuousness on the part of man might well arouse the wrath of the gods. "One should quench arrogance (hybris) rather than a conflagration" [38], says Heracleitus. Human laws, being concerned with maintaining the proper measure in man's relations with his fellows, punish presumptuous acts against the society in which he lives. In just the same way, the cosmic powers are constantly on guard lest any human being trespass against them. Science rationalized the mythological powers and turned them into laws of nature. But this did not alter the Greek's fundamental sense of being at the mercy of these powers and his horror of any meddling in their province, especially as the effect of the rationalization was neither deep nor lasting. We must remember that in the course of the Hellenistic period irrational tendencies reassumed the ascendancy, in consequence of the penetration of oriental culture into Greece and Rome. From the second century B.C. onwards there was a great spread of astrology, which originated in Babylon and Egypt. Helped by its inclusion in the teaching of the Stoic School, this took root in all the circles of Greek and Roman society, not excluding those of the educated. Medicine and alchemy were tainted by magic; and other irrational doctrines about mysterious virtues in the world of animal life, vegetation and inorganic matter were widely disseminated. All these tendencies had the effect of strengthening the inhibitions against technical undertakings and the bold practical application of science.

The general sense of insecurity resulting from political circumstances also had its effect in discouraging, rather than encouraging, technical inventions. Proof of this can be found in the military techniques developed in the third and second centuries by Archimedes, Ctesibius, Philo and Hero, who used the elasticity of taut ropes and compressed air in the construction of war-machines. These techniques, the child of urgent necessity, could have become the nucleus of considerable and many-sided technological developments, especially as the progress made in mathematics by Archimedes and his successors had laid the theoretical foundations for it. But in fact there is no sign of a development of this kind which might well have changed the economic structure of ancient society, or at least have left its imprint on one department of its economic life. Of the inventions of that period only Ctesibius' water-clock and water-organ deserve mention: all the others were merely show-pieces, such as Hero's siphon, or his instruments for demonstrating the motive power of steam. About all these technical inventions, even when they were accompanied by the construction of heavier mechanisms, there was something of a game or pastime. They were in fact toys, rather than means of harnessing the forces of nature for technical exploitation. Even though all the necessary preconditions existed, steam power was never exploited on a technical scale nor were the elements put to any economic use such as the construction of windmills or any but the most primitive type of water-mill.

The Ancient Greek believed fundamentally that the world should be *understood*, but that there was no need to *change* it. This remained the belief of subsequent generations up to the Renaissance. This passive attitude to the practical use of the forces of nature was reinforced by the complete ossification of the natural sciences in the Middle Ages in the condition to which Aristotle had brought them. The signs of a different attitude which appear here and there are no more than isolated sparks scattered through the darkness of centuries. A more economic and more varied use was made of streams and waterfalls as motive power for mills; windmills begin to be built in the twelfth century; the invention of gunpowder also gave an impetus to technical developments. But the real revolution, which in the first outburst of its pent-up forces completely transformed human affairs, came with the change of man's attitude to nature ushered in by the Renaissance. The Renaissance was the awakening of man's desire for conquest, the conquest and control of nature through science. Previously the attitude to nature had been one of submissiveness and the character of science had been theoretical and speculative. Now all this was thrust aside by the thirst for knowledge as the means of controlling the forces of nature and harnessing them to man's requirements. This striving for power through knowledge is one of the leading characteristics of the Renaissance: by understanding nature, the free man would be able to tame and exploit it for the extension of his own power. This revolution, which took place over a long period of time, cannot be traced to any single cause. But the significance of its consequences is plain enough: the fear of the gods and the elements was replaced by a spirit of adventurous conquest which turned science into the handmaid of technical progress. The supreme expression of this transformation of values is found in the personality of Leonardo da Vinci, the planner of harbours and canals, the dreamer of machines that would make man omnipotent, the technical visionary who, amongst other things, conceived the idea of the submarine and the aeroplane. This active, aggressive attitude of man to nature and his desire to take a hand in natural processes opened up a new world to him, while the increasingly complex structure of society directed this desire into the channel of scientific discovery and technical invention.

However, all this is still not sufficient to explain the peculiar quality of Greek science and its slow rate of progress. We must go back and examine its actual methods of acquiring knowledge in the light of our own. Functionally speaking, there is no difference at all: then, as now, the task of science was to systematize the sum total of our empirical knowledge in such a way as to make it possible to forecast future events. We have seen how ancient science performed this task most perfectly in astronomy. The empirical material accumulated from observations extending over centuries was worked into a deductive system. First of all, certain generalizations were made from which it was possible, for example, to establish a connection between the cycles of the planets and their distances from the earth—the swiftly revolving moon is the nearest, while the slowly revolving Saturn is the furthest away. Next came conjectures which brought uniformity into all the varied data. The hypothesis of spheres was developed, and after it the hypothesis of epicycles. These provided the astronomers with geometrical models which explained the known phenomena by a single principle: all the revolutions were reduced to circular movements. If the Greeks had discovered a new planet (the next planet, Uranus, was not discovered until 1781 by Herschel), they would straight away have been able to fit it into the existing framework and break down its movements by the established method. When the ever-increasing accuracy of observation led to the discovery of the precession of the equinoxes by Hipparchus, this too was added to the family of rotatory movements. When a still further increase in accuracy showed that some bodies diverge from a symmetrical orbit, the principle of circles was logically extended by the conjecture of eccentric circles. Here we have instances of that interaction of induction and deduction which also brought about the improvement of the calendar and made it possible to forecast solar and lunar eclipses more accurately. Thus scientific method had, as long ago as classical times and with no other means than geometrical and kinetic analogies, fully achieved in astronomy the functional task of science—that of fitting man *to foretell the future* with the greatest possible confidence. If we set the scientific maturity of Greek astronomy over against the feebleness of Greek achievements in "terrestrial" physics, we cannot help enquiring into the reason for this enormous contrast. The answer is to be found mainly in the great simplicity of astronomy when compared with the physical phenomena all round about us. As regards experimental conditions, the astronomic data are after all absolutely ideal. The objects of the experiment are points of light (or discs of light) whose movements are relatively simple; the framework of the fixed constellations permits a fairly accurate plotting of positions and changes of position, even with primitive observational apparatus; and, last but not least, the periodicity of the movements makes it possible to repeat the observations without limit after a certain time has elapsed. Here is a case where nature provides man with all the advantages of laboratory investigation, except for the possibility of arbitrarily changing the conditions. Paradoxical as it may seem, the secret of the simplicity of astronomy is to be found in the laboratory-like simplicity of the conditions under which it is studied. There is no parallel to these conditions in the natural phenomena of the earth. Here real progress only began to be made when man started to reproduce, albeit unconsciously, the convenient conditions of celestial observation in the laboratory experiment.

Systematic experimentation in the laboratory was first carried out in the seventeenth century. It was, to all intents and purposes, unknown to the Greeks. Thus the riddle of the backwardness of terrestrial physics in comparison with astronomy in antiquity resolves itself into the problem of why there was almost no use of the laboratory experiment in Greek science. The main characteristic of an experiment is its *artificiality*, which distinguishes it from the observation of a process in its natural form. In the realm of terrestrial physics observation is only the lowest stage of the experimental method. True, it is not to be despised, since in most cases it constitutes the starting-point for all subsequent investigation by experiment. But an experiment may also result from theoretical considerations by way of testing a given hypothesis, without any direct connection with any empirical observation whatsoever. The essential thing in an experiment is the isolation of a certain phenomenon in its pure form, for the purpose of studying it systematically. Herein lies its artificiality. Natural phenomena occur as part of a web of interwoven and interconnected processes; their continuity in time and space makes them appear to us as a single complete unit. Isolating a particular phenomenon from this unit is like an operation on a living body which separates one limb from the rest, or which places it in such a position that its functioning can be observed with as little interference as possible from the other parts. The classical example of this is provided by all those mechanical experiments in which friction or the resistance of environment is reduced as far as possible, so that the details of the pure mechanical process can be studied. It was on such experiments in the seventeenth century that the mechanics of Galileo and Newton were built. They were all based on the notion that friction or the resistance of environment are to be considered as incidental interferences with the study of the phenomenon that illustrates a natural law or principle in its pure form. This conception is as different as could be from Aristotle's. For him the environment was actually an integral part of the phenomenon itself, and he regarded the very idea of isolationas untenable. This point is particularly important, since the era of experimentation and the general revolution in the physical sciences started with the study of mechanical problems. Although it is unsafe to read a logical necessity into particular historical developments, the special position occupied by mechanics amongst the other branches of physics and natural science must be emphasized, for it was this special position that made it the starting-point of modern science. Most of the phenomena in man's surroundings, in the "anthropocentric" region, are of a mechanical nature and derive from mechanical forces, like gravity and elasticity. The action of our hands is also by direct mechanical contact. The fact that the physical experience of man is predominantly mechanical, determines to a large extent the character of his knowledge of the world around him which proceeds by mechanical analogies. Many physical concepts are borrowed from mechanics, as e.g. force, current, wave propagation and the simple atomic and molecular models. Notions like density and the whole kinetic terminology appear in electricity and optics and other parts of physics. It was, therefore, not by chance that mechanics headed the development of modern science and that the first mechanical experiments have had a decisive influence on the basic methods and notions of physics.

The artificiality of an experiment consists of more than the isolation of a phenomenon in the laboratory. If we wish to understand the full novelty of it and appreciate the gulf which separates it from the observation of a natural event, we must come down to the experimentation of our own day and see to what extremes of artificiality it has been carried. One of the important tasks of the experiment to-day is the confirmation of a given scientific theory. In many cases a theory is not capable of direct proof, but can be tested only by some conclusion which follows from its basic conceptions. The history of modern physics and chemistry teems with examples of this kind. It often happens that such conclusion, when translated into the terms of an experiment, does not correspond to any phenomenon actually found at any time or place in the material universe, either in the conditions of our earth or in any other astrophysical environment. The very idea of such an experiment, the way in which it is conducted, the apparatus set up for it and the processes revealed in the course of it are all the result of theoretical considerations. In carrying out this purely intellectual scheme the scientist produces a phenomenon which sometimes has no parallel in any natural process and the sole purpose of which is to confirm the given scientific theory. So we see that experiments of this kind are not intended to show how nature does function, but how it could function if the scientific conjecture prove to be correct. Here we have an extrapolation from actual to potential phenomena. The latter become actual only in the laboratory. In such a sense we may call an experiment unnatural. This, no doubt, is how it seemed to the Greeks, who would have thought it paradoxical to study natural phenomena by unnatural methods.

Modern science, therefore, has extended the conception of nature to include all phenomena whose existence is not ruled out by the laws governing the physical world. This development is an integral part of the conception of the cosmos as under the absolute law of causality which embraces both actual events and all those *possible* events that do not conflict with it. We have noted how the Greek mind was perplexed by the concept of the possible; even the Stoics, for all their progress in the causal category, ran into difficulties here. It was left for modern science to realize the distinction between what is *technically* impossible, and what is impossible *in principle*. The first is potentially possible within the framework of nature's laws, whereas the impossibility of the second is simply the negative expression of the existence of those laws. This distinction has not really been affected by the application of statistical laws to physics, however far-reaching the philosophical consequences of this may have been.

A direct use of the law of causality in the experimental method is *the repetition* of the experiment. Repetition obviously enables us to attain a greater degree of accuracy. But this is secondary to its main purpose, which is to confirm that a given phenomenon is governed by certain laws. Every time that we restore situation *A*, it gives rise to situation *B*. Thus the possibility of its repetition is one of the most important aspects of an experiment, and every repetition increases our certainty that the same cause always produces the same effect. This shows us once again the enormously important part played by the constantly recurring astronomical phenomena in forming the Greek scientific picture of the cosmos. In this respect the heavenly phenomena display all the ideal qualities of a laboratory experiment. The astronomer is again and again presented by nature with an unending sequence of the same initial conditions, the same situation *A* which gives rise to the same situation *B*. It was just the astronomical observation of these recurrences that awakened men's awareness of regularity in the cosmos, an awareness that was then further strengthened by observation of naturally recurring terrestrial phenomena, such as the rise and fall of tides. But, with few exceptions, it never occurred to the Greeks to devise *systematic* repetitions in imitation of nature, for the purpose of investigating the regularity of physical phenomena which are not self-repeating; they did not understand experiment as a series of identical, man-made events. Knowing of only natural, and not artificial, repetitions, they were unable to appreciate the great advantage of the latter over the former in the study of causality: in repeating an experiment we can change the initial conditions and test the effects of this change on the results, thus deepening our comprehension of causality. How will situation *B* vary as a function of a changing situation *A*? With this question and the answer given to it by experiment, the scientist in his laboratory passes from the single dimension of cause and effect to a multi-dimensional complex of interdependent causal possibilities. The systematic variation of the data of an experiment, with the study of the effect of this upon its course and outcome, is the complement of the experimental principle of isolating a phenomenon. These two processes—breaking up nature into isolated phenomena and repeatedly changing their course in an arbitrary way— have together speeded up our understanding of nature to an incredible extent. When set beside this breath-taking rate of advance, the progress of the Greeks in this field seems to us almost non-existent, based as it was on the study of things as they are in their entirety, but not as they may be when regarded as the sum totals of combinations of many factors. A slight development is discernible, as we have seen, in the systematic attempts of the military engineers in the Hellenistic period to improve ballistics. In this case there was a practical necessity for learning the connection between the efficient functioning of the machine and the size and shape of its various parts. Hence a more systematic study of technical problems took the place of haphazard gropings. This change of attitude is mentioned by both Hero and Philo. The development began with Archimedes. From his time onwards, we find occasional emphasis laid upon the continuity of scientific research and, side by side with explicit appeals to the authority of earlier investigators, hints for the guidance of future workers in the same field. Archimedes' words to this effect have already been quoted from his *Method* at the end of the last chapter. Philo of Byzantium, when discussing the size of openings through which pass the tension-producing elastic cables in ballistic machines, writes as follows: "Previous investigators failed to establish this size by their tests, since these were not conducted with various types of performance but only in connection with the required performance." (*Belopoeica*, 3.)

For all that, it must be reiterated that the Hellenistic scientists wrought no fundamental change and that the Greek aversion to experiment, especially the repetitive aspect of it, remained as strong as ever. Another indication of this, and one of great importance, is the absence of statistical calculation in antiquity. This has already claimed our attention in the chapter on "The Interdependence of Things". There we saw that the law of probability and the science of statistics developed out of the study of sequences in games of dice. These sequences provided an excellent opportunity for studying the recurrence of identical cases or similar combinations. The Greeks, though so attentive to the recurrent cycles of the heavens, showed no interest in the repetitions which occurred in the course of a human game of dice. As we know, it was not until the Renaissance that men began to examine problems of this kind mathematically. Cardano, in the middle of the sixteenth century, was the first (in his book on Dice Throwing) to raise the questions: What are all the possible results of throws with two dice? And how many times do combinations adding up to the same total appear in these throws? With the formulation of these questions he at once solved one of the problems of the law of probability, which was put on a sound methodical basis a hundred years later by Pascal. Pascal, too, started from problems presented by the games of chance of his day. This proves the decisive importance of the purposely repeated sequence for the understanding of regularity of nature. Only at a later stage in the analysis of such seemingly artificial sequences did there develop a theory of natural statistical events, with the discussion of the problems of the expectation of life and all the rules of the large numbers involved in phenomena of this kind. Insurance business also started in the Renaissance with the insurance of merchant ships bound for distant ports. Finally, in the seventeenth and eighteenth centuries, the mathematical theory of statistical regularity and probability was developed until "mathematical forecasting" became a branch of science. It may perhaps be argued that the Ancient Greek belief in Fate, placed upon a scientific basis by the Stoics, prevented him from recognizing the laws of chance. Such an argument is, however, superficial. The modern believer found an expression of divine providence in statistical causality no less than in the dynamic causality of Newtonian mechanics. In the introduction to his *Cosmogony*, Kant tries to show that the mathematical regularity of nature, so far from restricting God's authority, as the atheists maintain, is the most sublime manifestation of His infinite intelligence. A few years later (1761) a Prussian clergyman, Süssmilch, published a fundamental work of statistical research establishing the law of large numbers, the title of which indicates the author's own attitude to the question: *The Divine Order of Variations in Human Sex, as Proved by Births and Deaths and Natural Increase*. There is no reason for supposing that the Ancient Greek, had he discovered statistical regularity, would not have reconciled the religious and scientific aspects of the problem in a similar way. But he did not discover it, for the same reason that he did not develop systematic experimentation—because he was unable to transfer the idea of repetition from celestial to terrestrial occurrences; first from the "natural" to the "artificial" event, and then from the repetition of identical events to the repetition of combinations of events governed by more complex laws. Herein alike lies the explanation of there being no experimentation and no conception of probability in antiquity.

Just as the "dissection of nature" by experiment (to use Bacon's happy definition) was foreign to the Greek, so the corresponding theoretical process of describing nature in mathematical terms was also alien to his spirit. Once more we must turn to the beginning of the modern period to appreciate the decisive role in the rapid development of the natural sciences played by the application of mathematics to them. In antiquity, the use of mathematics in physical problems was confined to static phenomena where a mechanical question could easily be translated into geometrical or arithmetical terms, and to simple kinetic phenomena in which there exist simple relations between the distance covered and the time taken. It is true that although

Aristotle actually used the concept of velocity, as a relation of distance and time, he gives no precise mathematical definition. Accordingly, there is no hint of a suggestion of any quantitative definition of accelerated motion dependent upon the concept of acceleration, i.e. change of velocity with time. This revolutionary step was taken by Galileo, who developed the concepts of velocity and constant acceleration as part of his analysis of the laws of falling bodies in his book: *Mathematical Discourses and Proofs in Two New Branches of Science Concerning Mechanics and the Laws of Falling Bodies* (1638). Though using only elementary mathematics and basic theorems of proportion, Galileo gives explicit and clear definitions. After describing his experiment on falling bodies, he goes on to the exposition of the general case of accelerated motion either vertically or along an inclined plane. Galileo's work was revolutionary in two respects: in the actual evolving of formulae for accelerated motion; and in treating time as a *mathematical* quantity which could be used in calculations just like length or any other geometrical quantity. In his well-known theorems he uses such quantities as squares of time or squares of velocity and also roots of heights or the geometrical means of other lengths. His proofs are accompanied by graphs showing portions of time as sections on a straight line. This geometrical representation of time by Galileo was a step of first-class historical significance. Plato, in the *Timaeus*, had identified time with the periodic movements of the heavens, an identification expressive of the whole Greek conception of time: its eternity was equated with the eternal revolutions of the heavenly spheres. Aristotle, who gives a more general definition of time as "the number of the movement in relation to earlier and later", also regards circular motion as the most suitable description of time. He remarks that human affairs, too, constitute a kind of circular succession of events, and that all things that in the order of nature pass from creation to decay move in a circle. "Even time itself is regarded as a circle" [150]. Thus throughout antiquity the concept of time remained inseparable from the measurers of time—the clocks. Every clock, whether it is the earth revolving upon its axis (or the reflection of this revolution in the skies), or a water-clock, pendulum clock, etc., is a cyclic mechanism the functioning of which can be depicted in terms of a circular motion. As against this, the mathematical treatment

239

of the kinetic problems obliges us to disregard the essential difference between distance and time (as expressed in the fact that distances are measured only by measuring-rods and times by watches), and to comprehend them both in the abstract as co-ordinates, i.e. numerical quantities which are formally comparable. Physical time is simply a co-ordinate stretching from an arbitrary zero, fixed according to need, to infinity; its definition was an essential preliminary to the next step—comprehending locomotion as a function of time. We found the beginning of something like a functional conception amongst the Stoics. But they spoke of certain situations as a function of other situations; they did not get as far as seeing physical occurrences as functions of time comprehended as a geometrical dimension.

Galileo was the first to comprehend time in this way. This, together with the subsequent invention of the infinitesimal calculus, led the way to the complete mathematization of physics through the definition of physical quantities and their use in calculations. A physical quantity like velocity, acceleration, force, or such as the surface tension of a liquid, dielectrical constant, absorption coefficient, etc., is a mental abstraction: it is an artificial concept derived from experiment, capable of mathematical definition and serving as an instrument for the investigation of physical reality. The creation of a physical quantity starts with experience, frequently some specific experiment: acceleration, for example, was first studied in the case of falling bodies; then, expressed in the form of a general law, it became the instrument for analysing any and every phenomenon of dynamics. Generally speaking, the physical quantity was abstracted from experiences of various kinds which were first observed in relation to the human body, such as the power of human muscles. Then their results were enlarged by induction and deduction into general laws, until finally they assumed the form of a definition suitable for use over a very wide field of physical knowledge. Highly abstract though most physical quantities are, they are all rooted in experience, sometimes even in systematic experiments which, starting from theoretical calculations, have ended in the exhaustive mathematical summing-up of a given empirical fact. The physical quantities are the corner-stones of that whole edifice of mathematical physics wherein the sum total of our knowledge about the material universe is displayed in deductive form. It is thanks mainly to these quantities that mathematical equations dealing with problems of physics are distinguishable from pure mathematics; and it is because of them that the results of theoretical calculations can be retranslated into the language of experience. The physical quantity is thus obviously part of material reality, a part which plays a vital role in the process of understanding that reality. At the same time, it is just as obviously artificial in character, essentially a means to an end. It, too, is a kind of dissection performed by man upon nature to increase his knowledge of nature. In this sense it is the theoretical counterpart of the experiment. Clearly, then, the Ancient Greek would have found physical quantities and the whole mathe- matization of science absolutely unnatural. Movement and rest are natural phenomena; but velocity, as mathematically defined, is the relation between two such essentially different quantities as distance and time. The ways of measuring more complex physical quantities, i.e. the definitions given them in accordance with the combinations of the quantities which they contain, merely throw their artificiality into greater prominence. The concept of energy or work, which since the middle of the last century has come to dominate the descriptions of nature, is measured as mass multiplied by the square of velocity—a composite factor in which a Greek would have had difficulty in finding any connection at all with reality as he saw it.

Systematic experiment and the mathematization of natural science, which began simultaneously in the modern era, are part of the revolution which also brought technical development in its train. This development started, as we have seen, when man's attitude to nature became aggressive, when he was no longer content with just understanding nature, but was fired by the ambition to master it and the desire to exploit its forces for his own needs. "The dissection of nature" by experiment and mathe- matics was also the result of man's changed attitude to the cosmos. Greek science was born when the bonds which bound man to mythos were severed and he attached himself to logos. Nevertheless, this process of severance was never completed in antiquity; the Greek remained closely attached to the cosmos, as a result of his viewing the cosmos as a living organism, a body that can be understood and comprehended *in its entirety*. The Greek had a profound awareness of the unity of man and the cosmos, an awareness which was characterized by his biological approach to the world of matter. The teleological principle is essentially biological and anthropomorphic, so that the first basis for the conception of order in the cosmos was found in the system of the world of living things. Whereas we are reducing biology to physics and chemistry, the Greek applied the concepts and thought processes of biology to physical phenomena. Hence the contribution to physical thought made, directly and indirectly, by biologists and medical men from Hippocrates to Galen. The methods used in antiquity for studying the world of matter were thus similar to those employed in studying nature and living things, and the means employed were "natural", that is to say, based mainly on observation. The assumption of an absolute antithesis between heaven and earth, adopted by Greek science with only occasional deviations, was likewise a result of that subservience to myth which the Greek never succeeded in throwing off completely. We have seen how harmful this antithesis was to the advancement of science.

All these factors taken together resulted in the inability of the nation which created the natural sciences and methodical scientific thought to advance their development beyond the first stages. In the absence of experimentation and technical invention the process of scientific creation began to suffer from "shortage of fuel". The first signs of this appeared in the second century B.C. Its serious consequences were made still worse by the penetration of superstitions into the domain of science and the growth of occult tendencies consequent upon the merging of East and West in the Hellenistic era. This decline of creative science became part of the general eclipse of the Ancient World which followed on the disintegration of the Roman Empire and the collapse of political and civil security. With the spread of Christianity, natural problems only took second place to the main concern of humanity—its relations with the Creator. The petrification of science in the period of commentators and scholastics kept the knowledge of nature amassed by scientific research at the level attained by Aristotle until the beginning of the modern era. But this long period of immobility also brought about a slow, steady change in man's attitude to the cosmos. The last traces of the old Greek mythological subservience to the cosmos were eliminated by the influence of Christianity and the organized Church. By divorcing man and his vital interests from natural phenomena the Church helped to create the feeling that the cosmos was something alien and remote from man. It was this feeling that prepared men's minds for the next stage in which the investigator faced nature as its dissector and conqueror and thereby ushered in our own scientific era which still, after four centuries, retains its vigour undimmed.

Egyptian civilization created technology in the prescientific era. On its decline, the civilization of Greece gave birth to science without its technical application. Then, after a thousand years of paralysis, the civilization of Europe inaugurated the era of integration of science and technology. As participators in this era, we are in danger of suffering from a distorted perspective. Still, there can be no gainsaying the fundamental fact that this integration has become the mainspring of a creativeness and rapid progress, alike in the theoretical and practical spheres, which are unparalleled in earlier cultures. The fructification of technology by science is plain for all to see; the converse effect is no less profound and many sided. Not only does technology give fresh impetus to pure science, but technical achievements have been harnessed to the service of fundamental science. In this connection, it is sufficient to mention the tremendous service rendered by the development of scientific instruments and scientific machines in extending man's knowledge of nature beyond the limits of his five senses, thus enabling him to overcome that "weakness of the senses" which Anaxagoras regarded as the chief obstacle to ascertaining the truth. If the intellectual adventure of modern science is perhaps the greatest of all the adventures inaugurated by the modern era, this is due to the development of mathematics as the key to nature's laws. It is true that our cosmos has been drained of all the "human" content which it contained in the Greek period; it is true that the naïve world of the senses is separated from the world of science by an ever-widening chasm; it is true that understanding this world of science calls for enormous powers of abstraction and a professional and intellectual training which is becoming ever more rigorous. But, on the other hand, this cosmos—from the nucleus of the atom to the distant galaxies—is being filled more and more with new and marvellous contents which make the experience of those who share in this development certainly not less rich than the cosmic experience of the first natural philosophers of Ancient Greece.

That those same philosophers were among the spiritual ancestors of our own era will not be doubted by anyone who compares the heritage of Greek science—its methodical approach, the vigour of its imagination and inspiration, its associative strength and power of inference—with the science of our own time. Within the limits set to their cosmos by history, the Greeks, through their great spiritual resources, succeeded in weaving a marvellously rich and varied tapestry of thought which surprises us by its close resemblance to our own mental world. Within the limits of their scientific language they stated all the essential things that can be stated about the conformity to law of phenomena as regards their number and sequence, and about the interrelatedness of the various elements of physical reality. Their intuitive grasp of the atomic theory is such as to arouse our astonishment no less than their mathematical devices for explaining the movements of the heavenly bodies. The same inspiration is the source of their clearsighted vision in the framing of mechanistic cosmogonies at a time when the machine was still unknown, as also their penetrating analyses of various epistemological problems, such as the relation between the human senses and the human mind. Whoever makes a close study of the scientific world of Ancient Greece cannot but be filled with veneration and his veneration will but increase, the more he realizes that, beyond all differences and changes, the cosmos of the Greeks is still the rock from which our own cosmos has been hewn.

LIST OF SOURCES QUOTED

For texts translated from Diels, *Fragmente der Vorsokratiker* (6th ed.) and those translated from Arnim, *Stoicorum Veterum Fragmenta*, references to these collections are given in parentheses, marked with D and A respectively.

Thales (c. 624-546 B.C.)

1 Arist. *Metaph.* 983b
2 Herod. I 74
3 Arist. *de Caelo* 294a
4 Seneca, *natur. quaest.* III 14

Anaximander (c. 610-545 B.C.)

5 Simplicius, *Phys.* 24, 13 (D12A9)
6 *Strom.* 2 (D12A10)
7 Arist. *de Caelo* 295b
8 Agathemerus, I 1 (D12A6)
9 Diog. Laert. II 2 (D12A1)
10 Aët. II 20, 1; 24, 2 (D12A21)
11 Aët. II 13, 7 (D12A18)

Anaximenes (died c. 525 B.C.)

12 Simpl. *Phys.* 24, 10 (D13A5)
13 Aët. I, 3, 4 (D13B2)
14 Hippol. *Refut.* I 7 (D13A7)
15 *Strom.* 3 (D13A6)

Pythagoras and his School
(*middle of sixth century to middle of fourth century B.C.*)

16 Philolaus fragm. (D44B1)
17 ,, fragm. (D44B6)
18 ,, fragm. (D44B4)
19 ,, fragm. (D44B11)
20 ,, fragm. (D44B12)
21 Archytas fragm. (D44B1)
22 Arist. *Metaph.* 986a

23 Arist. *Phys.* 213b
24 Diog. VIII 48 (D28A44)
25 Theon Smyrn. 61, 11 (D47A19a)
26 Iambl. *in Nicom.* 100 (D18A15)
27 Nicom. *Arithm.* 26, 2 (D44A24)
28 Alex. Aphr. *Metaph.* 985b
29 Arist. *de Caelo* 290b
30 Arist. *Metaph.* 989b
31 Arist. *de An.* 405a
32 Clemens Alex. *Protr.* 66 (D24A12)
33 Arist. *de Caelo* 293a
34 Arist. *Metaph.* 986a
35 Cicero, *Acad. pr.* II 39 (D50, 1)
36 Eudemus, *Phys.* B III fragm. 51 (D58B34)

Heracleitus (c. 540-475 B.C.)

37 Heracleitus fragm. (D22B90)
38 Heracleitus fragm. (D22B43)
39 Arist. *de Caelo* 279b
40 Simpl. *de Caelo* 94, 4 (D22A10)

Empedocles (c. 500-430 B.C.)

41 Empedocles fragm. (D31B17)
42 ,, ,, (D31B16)
43 ,, ,, (D31B100)
44 ,, ,, (D31B13)

245

45 Arist. *Metaph.* 984a
46 Arist. *de Gen. et Corr.* 333a
47 Simpl. *Phys.* 25, 21 (D31A28)
48 Arist. *Phys.* 252a
49 Arist. *de Sensu* 446a
50 Philop. *de Anima* 344, 34 (D31A57)
51 Arist. *de Gen. et Corr.* 324b
52 Theophr. *de Sensu* 12
53 Plut. *quaest. nat.* 916D
54 Alex. Aphr. *quaest.* II 23 (D31A89)
55 Arist. *de Caelo* 295a

Anaxagoras (c. 488-428 B.C.)

56 Anaxagoras fragm. (D59B4)
57   ,,   ,,   (D59B12)
58   ,,   ,,   (D59B13)
59   ,,   ,,   (D59B16)
60   ,,   ,,   (D59B21)
61   ,,   ,,   (D59B21a)
62 Hippol. *Refut.* I 8 (D59A42)
63 Plut. *Lysand.* 12
64 Plut. *de fac. in orb. lun.* 929B
65 Diog. II 8 (D59A1)
66 Arist. *de Caelo* 295a
67 Arist. *Metaph.* 985a

Zeno of Elea (c. 450 B.C.)

68 Arist. *Phys.* 239b

Diogenes of Apollonia (c. 430 B.C.)

69 Diogenes fragm. (D64B3)

Leucippus (c. 450 B.C.)

70 Leucippus fragm. (D67B2)
71 Arist. *de Gen. et Corr.* 325a
72 Arist. *de Gen. et Corr.* 325a
73 Arist. *de Caelo* 276a
74 Arist. *de Gen. et Corr.* 314a
75 Simpl. *de Cael.* 242, 15 (D67A14)

76 Arist. *de Caelo* 303a
77 Aët. IV 9 (D67A32)
78 Aët. IV 13 (D67A29)
79 Alex. Aphr. *de Sensu* 24 (D67A29)
80 Arist. *Metaph.* 985b
81 Arist. *de Gen. et Corr.* 315b
82 Arist. *de Gen. et Corr.* 315a
83 Diog. IX 31 (D67A1)

Democritus (c. 460-370 B.C.)

84 Democritus fragm. (D68B6)
85   ,,   ,,   (D68B7)
86   ,,   ,,   (D68B8)
87   ,,   ,,   (D68B9)
88   ,,   ,,   (D68B117)
89   ,,   ,,   (D68B11)
90   ,,   ,,   (D68B164)
91   ,,   ,,   (D68B285)
92 Diog. IX 44 (D68A1)
93 Aët. I 16 (D68A48)
94 *Strom.* 7 (D68A39)
95 Simpl. *de Caelo* 294, 33 (D68A37)
96 Arist. *de Gen. et Corr.* 326a
97 Aët. I 3 (D68A47)
98 Aët. I 12 (D68A47)
99 Cicero, *de fato* 46
100 Galen, *de elem. sec. Hippocr.* I 2 (D68A49)
101 Theophr. *de sensu* 61-62
102   ,,   ,,   73-75
103   ,,   ,,   65-67
104   ,,   *de caus. plant.* VI 7, 2
105 Aët. III 1 (D68A91)
106 Achill. *Isag.* 24 (D68A91)
107 Galen, *de medic. empir.* 1259, 8 (D68B125)
108 Sext. Emp. *adv. math.* VIII 6 (D68A59)
109 *Strom.* 7 (D68A39)
110 Arist. *Phys.* 252a

111 Aët. I 29 (D59A66)
112 Diog. IX 44 (D68A1)
113 Arist. *Phys.* 196a
114 Hipp. *Refut.* I 13 (D68A40)

Eudoxus (c. 409-356 B.C.)

115 Simpl. *de Caelo* 498a
116 Arist. *Metaph.* 1073b

Heracleides (c. 388-315 B.C.)

117 Chalcid. *in Tim.* 110
118 Mart. Cap. *de nupt. Merc. et Phil.* VIII
119 Aët. III 13 (D51, 1)

Plato (429-347 B.C.)

120 Plato, *Phaedo* 92D
121 ,, ,, 97B
122 ,, *Theaet.* 147D
123 ,, ,, 162E
124 ,, *Phileb.* 55D
125 ,, *Symp.* 175D
126 ,, *Phaedr.* 245C
127 ,, *Io* 533D
128 ,, *Republ.* 529A
129 ,, ,, 590C
130 ,, *Tim.* 46D
131 ,, ,, 47E
132 ,, *Laws* 886D
133 ,, ,, 897C
134 ,, ,, 898A
135 ,, ,, 967A
136 ,, *Epinom.* 982C
137 Theophr. *de sensu* 88
138 Plut. *quaest. Plat.* 1006C

Aristotle (384-322 B.C.)

139 *Anal. Pr.* 41a
140 *Anal. Post.* 95b
141 *Phys.* 198b
142 ,, 214b

143 *Phys.* 215a
144 ,, 215b
145 ,, 216a
146 ,, 219a
147 ,, 219b
148 ,, 220b
149 ,, 223a
150 ,, 223b
151 ,, 249b
152 ,, 265a
153 *de Caelo* 269a, 270b
154 ,, ,, 272b
155 ,, ,, 273b
156 ,, ,, 277b
157 ,, ,, 289a
158 ,, ,, 290a
159 ,, ,, 292a
160 ,, ,, 296b
161 ,, ,, 297b
162 ,, ,, 300b
163 ,, ,, 307b
164 ,, ,, 309b
165 *de Gen. et Corr.* 327b
166 *Meteor.* 339a
167 *Metaph.* 980a
168 ,, 982b
169 ,, 1061a
170 ,, 1071b
171 ,, 1073b
172 Simpl. *Phys.* 325, 24
173 Cicero, *de nat. deor.* II 16

Theophrastus (372-287 B.C.)

174 Theophr. *Metaphys.* 1
175 ,, ,, 27

Epicurus (341-270 B.C.)

176 *Letter to Herod.* 41
177 ,, ,, 46
178 ,, ,, 50
179 ,, ,, 54
180 ,, ,, 56
181 ,, ,, 61

182 *Letter to Herod.* 76
183 *Letter to Pyth.* 85-87
184   ,,   ,,   88
185   ,,   ,,   91-97
186 *Letter to Menoec.* 134
187 *Principal Doctrines* 23

Zeno of Cition (*c.* 332-262 B.C.)

188 Cicero, *de nat. deor.* II 57
189   ,,   ,,   ,, II 23
190 Euseb. *praep. evang.* XV
       (AI98)
191 Stob. *Eclog.* I 171 (A II 596)

Chrysippus (*c.* 280-207 B.C.)

192 Galen, *de anim. mor.* IV 783
       K (A II 787)
193 Alex. Aphr. *de mixt.* 216, 14
       (A II 473)
194 Galen, *de plenitud.* 3 (A II
       439)
195 Galen, *de plenitud.* 3 (A II
       440)
196 Plut. *de comm. not.* 1085D
197 Plut. *de Stoic. repugn.* 1053F
198 Alex. Aphr. *de mixt.* 224, 14
       (A II 442)
199 Aët. IV 19 (A II 425)
200 Diog. VII 158 (A II 872)
201 Diog. VII 157 (A II 867)
202 Nemes. *de nat. hom.* 2, 42
       (A II 451)
203 Philo, *Quod deus sit immut.*
       35 (A II 458)
204 Philo, *de sacrif. Abel et Cain*
       68 (A II 453)
205 Galen, *de muscul. motu* I 7-8
       (A II 450)
206 Diog. VII 140 (A II 543)
207 Stob. *Eclog.* I 153 (A II 471)
208 Alex. Aphr. *de mixt.* 216, 14
       (A II 473)

209 Plut. *de comm. not.* 1078E
210 Alex. Aphr. *de anima*, 140,
       10 (A II 477)
211 Stob. *Eclog.* I 106 (A II 509)
212 Plut. *de comm. not.* 1081F
213   ,,   ,,   ,, 1081C
214   ,,   ,,   ,, 1082A
215   ,,   ,,   ,, 1079F
216   ,,   ,,   ,, 1080C
217   ,,   ,,   ,, 1078E
218   ,,   ,,   ,, 1079A
219 Plut. *de Stoic. repugn.* 1045C
220 Alex. Aphr. *de fato* 22
       (A II 945)
221 Stob. *Eclog.* I 79 (A II 913)
222 Gellius, *noct. att.* VII 2
       (A II 1000)
223 Cicero, *de fato* 41
224 Gellius, *noct. att.* VII 2
       (A II 1000)
225 Cicero, *de divin.* I ˙09
226   ,,   ,, I 118
227 Euseb. *praep. evang.* IV
       (A II 939)
228 Plut. *de Stoic. repugn.* 1055D
229 Alex. Aphr. *de fato* 10
       (A II 959)
230 Diog. VII 151 (A II 693)
231 Plut. *de Stoic. repugn.* 1053A
232 Alex. Aphr. *in meteor.* 90a
       (A II 594)
233 Alex. Aphr. *in anal. pr.* 180,
       31 (A II 624)
234 Lactant. *div. instit.* VII 23
       (A II 623)

Aristarchus (*c.* 310-230 B.C.)

235 Plut. *de fac. in orb. lun.* 923A
236 Aristarch. fragm.

Archimedes (287-212 B.C.)

237 Archim. *Sand-reckoner* I

Hipparchus (c. 190-120 B.C.)

238 Pliny, *nat. hist.* II 26, 95
239 Ptolemy, *Synt. math.* VII 2

Poseidonius (c. 135-51 B.C.)

240 Aët. II 25
241 Cleom. *de motu circul. doctr.*
    I 10, 50
242 Cleom. *de motu circul. doctr.*
    I 1, 4
243 Cleom. *de motu circul. doctr.*
    I 1, 6
244 Strabo, *geogr.* III 5

Lucretius (c. 95-55 B.C.)

245 Lucr. *de rer. nat.* I 265-270,
    298-321
246 Lucr. *de rer. nat.* II 109
247 Lucr. *de rer. nat.* II 112-141
248 Lucr. *de rer. nat.* II 217-260,
    277-280, 284-293
249 Lucr. *de rer. nat.* II 308-322
250 Lucr. *de rer. nat.* II 757-771
251 Lucr. *de rer. nat.* II 1007-
    1022
252 Lucr. *de rer. nat.* IV 478-
    485; 499

253 Lucr. *de rer. nat.* V 235-246
254 Lucr. *de rer. nat.* V 419-431

Strabo (born c. 63 B.C.)

255 Strabo, *geogr.* I 1

Seneca (c. 3 B.C.-A.D. 65)

256 Seneca, *quaest. nat.* VII 17
257 ,,   ,,   ,, VII 25

Plutarch (c. 46-120 A.D.)

258 Plut. *de fac. in orb. lun.* 923D
259 ,,   ,,   ,,   924A
260 ,,   ,,   ,,   924E
261 ,,   ,,   ,,   925F
262 ,,   ,,   ,,   926E
263 ,,   ,,   ,,   927C
264 ,,   ,,   ,,   927E
265 ,,   ,,   ,,   929E
266 ,,   ,,   ,, 930A, D
267 ,,   ,,   ,,   931C
268 ,,   ,,   ,,   935C
269 ,,   ,,   ,, 936B, D
270 ,,   ,,   ,,   938D

Ptolemy (c. 150 A.D.)

271 Ptolemy, *Synt. math.* I 7

SELECTED BIBLIOGRAPHY

ARNIM, J. v., *Stoicorum Veterum Fragmenta*, Leipzig, 1903.

BAILEY, C., *The Greek Atomists and Epicurus*, Oxford University Press, 1928.

BRÉHIER, E., *Chrysippe*, Presses Universitaires de France, Paris, 1951.

BURNET, J., *Early Greek Philosophy*, 4th edit., Black, London, 1952.

CHERNISS, H., *Aristotle's Criticism of Presocratic Philosophy*, Johns Hopkins University Press, 1935.

COHEN, M. R., and DRABKIN, J. E., *A Source Book in Greek Science*, McGraw-Hill, New York, 1948.

CORNFORD, F. M., *Plato's Cosmology*, Routledge & Kegan Paul, London, 1937.

CORNFORD, F. M., *Principium Sapientiae*, Cambridge University Press, 1952.

DIELS, H., *Die Fragmente der Vorsokratiker*, 6th edit. by W. Kranz, Berlin, 1951.

DIELS, H., *Antike Technik*, 3rd edit., Leipzig, 1924.

DODDS, E. R., *The Greeks and the Irrational*, University of California Press, Berkeley, 1951.

DREYER, J. L. E., *A History of Astronomy from Thales to Kepler*, 2nd edit., Dover Publications Inc., 1953.

DUHEM, P., *Le Système du Monde*, Vol. I and II, Paris, 1913.

EDELSTEIN, L., "Recent Trends in the Interpretation of Ancient Science", *Journal of the History of Ideas*, XIII, pp. 573-604.

FARRINGTON, B., *Greek Science*, Penguin Books, Vol. I, 1944; Vol. II, 1949.

HEATH, T., *Greek Astronomy*, Dent, London, 1932.

JAEGER, W., *Aristotle*, transl. by R. Robinson, 2nd edit., Oxford University Press, 1948.

MATES, B., *Stoic Logic*, University of California Press, Berkeley, 1953.

POHLENZ, M., *Die Stoa*, Vandenhoeck and Ruprecht, Goettingen, 1948.

SARTON, G., *Introduction to the History of Science*, Williams and Wilkins, Baltimore, 1927-48.

SCHRÖDINGER, E., *Nature and the Greeks*, Cambridge University Press, 1954

INDEX

References to quotations are printed in heavy type

251

DATE DUE